Big Ideas in Mathematics for Future Mathematics Teachers

Big Ideas in Infinite Processes

John Beam, Jason Belnap, Eric Kuennen,
Amy Parrott, and Jennifer Szydlik
(Updated Spring 2021)

Copyright 2021 by John Beam, Jason Belnap, Eric Kuennen, Amy Parrott, and Jennifer Szydlik

This work is licensed under the Creative Commons Attribution-NonCommercial-NoDerivatives 4.0 International License. To view a copy of this license, visit http://creativecommons.org/licenses/by-nc-nd/4.0/ or send a letter to Creative Commons, PO Box 1866, Mountain View, CA 94042, USA.

Cover photo credit: Max Ronnersjö/Alexander S. Peak/Wikimedia Commons/CC-BY-SA-3.0

ISBN: 9798525819481

Eric W. Kuennen, Mathematics Department, University of Wisconsin Oshkosh
800 Algoma Blvd, Oshkosh, WI 54901

Dear Future Teacher,

We wrote this book to help you to see the structure that underlies elementary and middle school mathematics, to give you experiences really *doing* mathematics, and to show you how children think and learn. We fully intend this course to transform your relationship with math.

As teachers of future math teachers, we created or gathered the activities for this text, and then we tried them out with our own students and modified them based on their suggestions and insights. We know that some of the problems are tough – you will get stuck sometimes. Please don't let that discourage you. There's much value in wrestling with an idea.

All our best,

John, Jason, Eric, Amy, Carol & Jen

Table of Contents

The infinite! No other question has ever moved so profoundly the spirit of man.
David Hilbert
The World of Mathematics

HEY! READ THIS. IT WILL HELP YOU UNDERSTAND THE BOOK.	**1**
COMMON CORE STATE STANDARDS FOR MIDDLE SCHOOL ALGEBRA	3
COMMON CORE STATE STANDARDS FOR MATHEMATICAL PRACTICE	7

Chapter One: Infinity

CLASS ACTIVITY 1: THE HAUNTED HOUSE	**11**
APPARENT PARADOXES, ACHILLES AND THE TORTOISE	12
CARDINALITY	14
SIEVE OF ERATOSTHENE	15
CLASS ACTIVITY 2: SO MANY PRIMES!	**16**
FUNDAMENTAL THEOREM OF ARITHMETIC	16
INFINITUDE OF PRIMES: A FIRST PROOF BY CONTRADICTION	16
LANGUAGE OF MATHEMATICS	18
CLASS ACTIVITY 3: EXPLORING INFINITE LISTS	**22**
AN INTRODUCTION TO SEQUENCES	23
FORMAL DEFINITION OF CONVERGENCE	23
CLASS ACTIVITY 4: EXPLORING INFINITE SUMS	**29**
AN INTRODUCTION TO SERIES	30
DIVERGENCE AND CONVERGENCE	30
CLASS ACTIVITY 5: RATIONAL NUMBERS AND GEOMETRIC SERIES	**35**
DECIMAL REPRESENTATIONS OF REAL NUMBERS	35
GEOMETRIC SERIES	37
CLASS ACTIVITY 6: AN IRRATIONAL NUMBER	**40**
DENSE SETS	42

CLASS ACTIVITY 7: LIFE IN HELL — 46
- INFINITE SETS — 47
- COUNTABLE AND UNCOUNTABLE SETS — 48
- CANTOR'S DIAGONAL ARGUMENT — 49

CLASS ACTIVITY 8: TO INFINITY AND BEYOND! — 52
- POWER SETS — 52
- CARDINALITY OF INFINITE SETS — 53
- CANTOR'S THEOREM — 53

Chapter Two: Functions and Modeling

CLASS ACTIVITY 9: MAKING MODELS — 60
- FUNCTIONS AS MODELS — 61
- DOMAIN AND RANGE — 61

CLASS ACTIVITY 10: TIDES, RABBITS, AND FALLING OBJECTS — 69
- UNDERSTANDING NUMERIC DATA — 71
- LINEAR, QUADRATIC, TRIGONOMETRIC, AND EXPONENTIAL FORMS — 72

CLASS ACTIVITY 11: PICTURING FUNCTIONS — 77
- POLYNOMIALS — 78
- FUNDAMENTAL THEOREM OF ALGEBRA — 78

CLASS ACTIVITY 12: TAKE IT TO THE LIMIT — 83
- THE IDEA OF A LIMIT — 85
- CONTINUITY — 87
- CLASSIFICATION OF DISCONTINUITIES — 87

CLASS ACTIVITY 13: GROWING ALL THE TIME — 92
- THE DEFINITION OF e — 93
- LIMITS AT INFINITY — 94
- END BEHAVIOR AND ASYMPTOTES — 94

CLASS ACTIVITY 14: CIRCLE GETS THE SQUARE — 98
- ARCHIMEDES'S METHOD FOR ESTIMATING π — 99

Chapter Three: The Derivative

CLASS ACTIVITY 15: WALKING THE LINE — 102

CLASS ACTIVITY 15B: TOYING AROUND — 104
Graphing Position and Velocity — 104
Introduction to Calculus — 109

CLASS ACTIVITY 16: BASEBALL VELOCITY — 110
Definition of Derivative — 113

CLASS ACTIVITY 17: A DERIVATIVE SHORTCUT — 118
Leibniz's Notation — 119
The Power Rule — 120
Sum and Constand Multiple Rules — 121

CLASS ACTIVITY 18: BASEBALL ACCELERATION — 124
Second Derivative — 126
Concavity — 128

CLASS ACTIVITY 19: DERIVATIVES OF SINE, COSINE, AND e^x — 130
Derivatives of some Special Functions — 131
Antiderivatives — 132

CLASS ACTIVITY 20: GAS MILEAGE — 135
Function Composition — 137
Chain Rule for Derivatives — 138

CLASS ACTIVITY 21: POPCORN BOXES — 141
Extreme Values — 142
Optimization — 142
Differentiability — 143

Chapter Four: The Definite Integral

CLASS ACTIVITY 22: SPEED RACERS — 147
- Left and Right Hand Sums — 149
- The Definite Integral — 150

CLASS ACTIVITY 23: THE FUNDAMENTAL THEOREM OF CALCULUS — 154
- The Big Ideas of Calculus — 157
- Average Value of a Function — 160

CLASS ACTIVITY 24: A PIPELINE PROBLEM — 161
- An Optimization Project — 161

REFERENCES — 162

GLOSSARY: — 164

Hey! Read this. It will help you understand the book.

> *The last thing one knows when writing a book is what to put first.*
> Blaise Pascal, Pensees

This book was written to prepare future teachers for the mathematical work of teaching. The focus of this module is infinite processes and, in case you haven't figured it out yet, this means we are going to eventually get to the big ideas of calculus. We know that right about now you're thinking, *Calculus? I'm not going to be teaching calculus to my seventh grade class*. Okay, probably not. But the types of ideas that lead to calculus and the kind of thinking required will be a part of your middle grades curriculum, and we have the examples to prove it. The ideas in this book are fundamentally important for your students to understand and so they are fundamentally important for *you* to understand. In addition, the ideas of calculus are truly stunning and beautiful. Just you wait and see.

As mathematicians, we will also try to convey to you the beauty of our subject. Mathematicians view mathematics as the study of patterns and structures. We want to show you how to reason like a mathematician – and we want you to show this to your students too. This *way of reasoning* is just as important as any content you teach. When you stand before your class, you are a representative of the mathematical community; we will help you to become a good one.

No one can do this thinking for you. Mathematics isn't a subject you can memorize; it is about ways of thinking and knowing. *You* need to do examples, gather data, look for patterns, experiment, draw pictures, think, try again, make arguments, and think some more. The big ideas of probability and statistics are not always easy.

Each section of this book begins with a **Class Activity**. The activity is designed for small-group work in class. Some activities may take your class as little as 30 minutes to complete and discuss. Others may take you two or more class periods. No solutions are provided to activities – you will have to solve them yourselves. The **Read and Study**, **Connections to Teaching**, and **Homework** sections are presented within the context of the activity ideas.

> *One of the big misapprehensions about mathematics that we perpetuate in our classrooms is that the teacher always seems to know the answer to any problem that is discussed. This gives students the idea that there is a book somewhere with all the right answers to all of the interesting questions, and that teachers know those answers. And if one could get hold of the book, one would have everything settled. That's so unlike the true nature of mathematics.*
> Leon Henkin

To prepare ourselves to write this text we studied four *Standards*-based curriculum projects for middle school students (the books your future students might use). Those projects are

Mathematics in Context, *Connected Mathematics*, *MATHematics*, and *MathScape*. All of these are activity-based curricula. This means that the middle school materials were written so that your future students will solve problems and create understandings based on concrete experiences.

In case you are skeptical about these types of materials for your future students, let us assure you that they better encourage and support the types of behaviors and thinking that mathematicians value than do traditional materials. Furthermore, the research suggests that schools that had adopted *Standards*-based materials for more than two years showed significantly higher test scores on even traditional measures of mathematical understanding than did matched schools that adopted traditional curricula (Reys, Reys, Lapan, Holliday, & Wasman, 2003; Riordan & Noyce, 2001; Griffen, Evans, Timms, & Trowell, 2000). We assure you that the ideas you will meet in these pages are vitally connected to the mathematics curriculum of your future students, and we hope that the text is written in a way that makes these connections apparent to you.

The Common Core State Standards is "a state-led effort to establish a shared set of clear educational standards for English language arts and mathematics that states can voluntarily adopt. The standards have been informed by the best available evidence and the highest state standards across the country and globe and designed by a diverse group of teachers, experts, parents, and school administrators…" (see http://www.corestandards.org/Math/) As of the time of publication of this text, most states had officially adopted these standards, and so it is important for you to know them and the content and practices that they advocate.

The mathematics content in this book is focus on preparing you to teach the Common Core State Standards for Mathematics for grades 6 - 8. These are the standards that you will likely follow when you are a teacher, so we will highlight aspects of them throughout the text. In order for you to see how the mathematical work you are doing appears in the elementary grades, we have made explicit connections to *Core Connections* from College Preparatory Mathematics (CPM). This is the middle grades mathematics curriculum adopted by the Oshkosh Area School District. You will often be asked to go to the site https://cpm.org/university to read or do problems. Your instructor will provide you with a code so that you can access these materials.

On the next several pages are the Common Core State Standards for the Content Domains of Expressions & Equations and Functions in grades 6-8, as well as the standards for Mathematical Practice. Throughout the text, we will ask you to refer back to these standards to reflect upon.

Common Core State Standards for Middle School Algebra

Expressions and Equations

Grade 6
Apply and extend previous understandings of arithmetic to algebraic expressions.

1. Write and evaluate numerical expressions involving whole-number exponents.

2. Write, read, and evaluate expressions in which letters stand for numbers.

    a. Write expressions that record operations with numbers and with letters standing for numbers. *For example, express the calculation "Subtract y from 5" as $5 - y$.*

    b. Identify parts of an expression using mathematical terms (sum, term, product, factor, quotient, coefficient); view one or more parts of an expression as a single entity. *For example, describe the expression $2(8+7)$ as a product of two factors; view $(8 + 7)$ as both a single entity and a sum of two terms.*

    c. Evaluate expressions at specific values of their variables. Include expressions that arise from formulas used in real-world problems. Perform arithmetic operations, including those involving whole-number exponents, in the conventional order when there are no parentheses to specify a particular order (Order of Operations). *For example, use the formulas $V = s^3$ and $A = 6s^2$ to find the volume and surface area of a cube with sides of length $s = 1/2$.*

3. Apply the properties of operations to generate equivalent expressions. *For example, apply the distributive property to the expression $3(2 + x)$ to produce the equivalent expression $6 + 3x$; apply the distributive property to the expression $24x + 18y$ to produce the equivalent expression $6(4x + 3y)$; apply properties of operations to $y + y + y$ to produce the equivalent expression $3y$.*

4. Identify when two expressions are equivalent (i.e., when the two expressions name the same number regardless of which value is substituted into them). *For example, the expressions $y + y + y$ and $3y$ are equivalent because they name the same number regardless of which number y stands for.*

Reason about and solve one-variable equations and inequalities.

5. Understand solving an equation or inequality as a process of answering a question: which values from a specified set, if any, make the equation or inequality true? Use substitution to determine whether a given number in a specified set makes an equation or inequality true.

6. Use variables to represent numbers and write expressions when solving a real-world or mathematical problem; understand that a variable can represent an unknown number, or, depending on the purpose at hand, any number in a specified set.

7. Solve real-world and mathematical problems by writing and solving equations of the form $x + p = q$ and $px = q$ for cases in which p, q and x are all nonnegative rational numbers.

8. Write an inequality of the form $x > c$ or $x < c$ to represent a constraint or condition in a real-world or mathematical problem. Recognize that inequalities of the form $x > c$ or $x < c$ have infinitely many solutions; represent solutions of such inequalities on number line diagrams.

Represent and analyze quantitative relationships between dependent and independent variables.

9. Use variables to represent two quantities in a real-world problem that change in relationship to one another; write an equation to express one quantity, thought of as the dependent variable, in terms of the other quantity, thought of as the independent variable. Analyze the relationship between the dependent and independent

variables using graphs and tables, and relate these to the equation. *For example, in a problem involving motion at constant speed, list and graph ordered pairs of distances and times, and write the equation $d = 65t$ to represent the relationship between distance and time.*

Grade 7
Use properties of operations to generate equivalent expressions.

1 Apply properties of operations as strategies to add, subtract, factor, and expand linear expressions with rational coefficients.

2 Understand that rewriting an expression in different forms in a problem context can shed light on the problem and how the quantities in it are related. *For example, $a + 0.05a = 1.05a$ means that "increase by 5%" is the same as "multiply by 1.05."*

Solve real-life and mathematical problems using numerical and algebraic expressions and equations.

3 Solve multi-step real-life and mathematical problems posed with positive and negative rational numbers in any form (whole numbers, fractions, and decimals), using tools strategically. Apply properties of operations to calculate with numbers in any form; convert between forms as appropriate; and assess the reasonableness of answers using mental computation and estimation strategies. *For example: If a woman making $25 an hour gets a 10% raise, she will make an additional $\frac{1}{10}$ of her salary an hour, or $2.50, for a new salary of $27.50. If you want to place a towel bar $9\frac{3}{4}$ inches long in the center of a door that is $27\frac{1}{2}$ inches wide, you will need to place the bar about 9 inches from each edge; this estimate can be used as a check on the exact computation.*

4 Use variables to represent quantities in a real-world or mathematical problem, and construct simple equations and inequalities to solve problems by reasoning about the quantities.

   a Solve word problems leading to equations of the form $px + q = r$ and $p(x + q) = r$, where p, q, and r are specific rational numbers. Solve equations of these forms fluently. Compare an algebraic solution to an arithmetic solution, identifying the sequence of the operations used in each approach. *For example, the perimeter of a rectangle is 54 cm. Its length is 6 cm. What is its width?*

   b Solve word problems leading to inequalities of the form $px+q > r$ or $px+q < r$, where p, q, and r are specific rational numbers. Graph the solution set of the inequality and interpret it in the context of the problem. *For example: As a salesperson, you are paid $50 per week plus $3 per sale. This week you want your pay to be at least $100. Write an inequality for the number of sales you need to make, and describe the solutions.*

Grade 8
Work with radicals and integer exponents.

1 Know and apply the properties of integer exponents to generate equivalent numerical expressions. *For example, $3^2 \times 3^{-5} = 3^{-3} = 1/3^3 = 1/27$.*

2 Use square root and cube root symbols to represent solutions to equations of the form $x^2 = p$ and $x^3 = p$, where p is a positive rational number. Evaluate square roots of small perfect squares and cube roots of small perfect cubes. Know that $\sqrt{2}$ is irrational.

3 Use numbers expressed in the form of a single digit times a whole-number power of 10 to estimate very large or very small quantities, and to express how many times as much one is than the other. *For example, estimate the population of the United States as 3×10^8 and the population of the world as 7×10^9, and determine that the world population is more than 20 times larger.*

Copyright 2010. National Governors Association Center for Best Practices and Council of Chief State School Officers. All rights reserved.

4. Perform operations with numbers expressed in scientific notation, including problems where both decimal and scientific notation are used. Use scientific notation and choose units of appropriate size for measurements of very large or very small quantities (e.g., use millimeters per year for seafloor spreading). Interpret scientific notation that has been generated by technology.

Understand the connections between proportional relationships, lines, and linear equations.

5. Graph proportional relationships, interpreting the unit rate as the slope of the graph. Compare two different proportional relationships represented in different ways. *For example, compare a distance-time graph to a distance-time equation to determine which of two moving objects has greater speed.*

6. Use similar triangles to explain why the slope m is the same between any two distinct points on a non-vertical line in the coordinate plane; derive the equation $y = mx$ for a line through the origin and the equation $y = mx + b$ for a line intercepting the vertical axis at b.

Analyze and solve linear equations and pairs of simultaneous linear equations.

7. Solve linear equations in one variable.

   a. Give examples of linear equations in one variable with one solution, infinitely many solutions, or no solutions. Show which of these possibilities is the case by successively transforming the given equation into simpler forms, until an equivalent equation of the form $x = a$, $a = a$, or $a = b$ results (where a and b are different numbers).

   b. Solve linear equations with rational number coefficients, including equations whose solutions require expanding expressions using the distributive property and collecting like terms.

8. Analyze and solve pairs of simultaneous linear equations.

   a. Understand that solutions to a system of two linear equations in two variables correspond to points of intersection of their graphs, because points of intersection satisfy both equations simultaneously.

   b. Solve systems of two linear equations in two variables algebraically, and estimate solutions by graphing the equations. Solve simple cases by inspection. *For example, $3x + 2y = 5$ and $3x + 2y = 6$ have no solution because $3x + 2y$ cannot simultaneously be 5 and 6.*

   c. Solve real-world and mathematical problems leading to two linear equations in two variables. *For example, given coordinates for two pairs of points, determine whether the line through the first pair of points intersects the line through the second pair.*

Copyright 2010. National Governors Association Center for Best Practices and Council of Chief State School Officers. All rights reserved.

Functions

Grade 8

Define, evaluate, and compare functions.

1. Understand that a function is a rule that assigns to each input exactly one output. The graph of a function is the set of ordered pairs consisting of an input and the corresponding output.[1]

2. Compare properties of two functions each represented in a different way (algebraically, graphically, numerically in tables, or by verbal descriptions). *For example, given a linear function represented by a table of values and a linear function represented by an algebraic expression, determine which function has the greater rate of change.*

3. Interpret the equation $y = mx + b$ as defining a linear function, whose graph is a straight line; give examples of functions that are not linear. *For example, the function $A = s^2$ giving the area of a square as a function of its side length is not linear because its graph contains the points $(1, 1)$, $(2, 4)$ and $(3, 9)$, which are not on a straight line.*

Use functions to model relationships between quantities.

4. Construct a function to model a linear relationship between two quantities. Determine the rate of change and initial value of the function from a description of a relationship or from two (x, y) values, including reading these from a table or from a graph. Interpret the rate of change and initial value of a linear function in terms of the situation it models, and in terms of its graph or a table of values.

5. Describe qualitatively the functional relationship between two quantities by analyzing a graph (e.g., where the function is increasing or decreasing, linear or nonlinear). Sketch a graph that exhibits the qualitative features of a function that has been described verbally.

Common Core State Standards for Mathematical Practice

Mathematics | Standards for Mathematical Practice

The Standards for Mathematical Practice describe varieties of expertise that mathematics educators at all levels should seek to develop in their students. These practices rest on important "processes and proficiencies" with longstanding importance in mathematics education. The first of these are the NCTM process standards of problem solving, reasoning and proof, communication, representation, and connections. The second are the strands of mathematical proficiency specified in the National Research Council's report *Adding It Up*: adaptive reasoning, strategic competence, conceptual understanding (comprehension of mathematical concepts, operations and relations), procedural fluency (skill in carrying out procedures flexibly, accurately, efficiently and appropriately), and productive disposition (habitual inclination to see mathematics as sensible, useful, and worthwhile, coupled with a belief in diligence and one's own efficacy).

1 Make sense of problems and persevere in solving them.

Mathematically proficient students start by explaining to themselves the meaning of a problem and looking for entry points to its solution. They analyze givens, constraints, relationships, and goals. They make conjectures about the form and meaning of the solution and plan a solution pathway rather than simply jumping into a solution attempt. They consider analogous problems, and try special cases and simpler forms of the original problem in order to gain insight into its solution. They monitor and evaluate their progress and change course if necessary. Older students might, depending on the context of the problem, transform algebraic expressions or change the viewing window on their graphing calculator to get the information they need. Mathematically proficient students can explain correspondences between equations, verbal descriptions, tables, and graphs or draw diagrams of important features and relationships, graph data, and search for regularity or trends. Younger students might rely on using concrete objects or pictures to help conceptualize and solve a problem. Mathematically proficient students check their answers to problems using a different method, and they continually ask themselves, "Does this make sense?" They can understand the approaches of others to solving complex problems and identify correspondences between different approaches.

2 Reason abstractly and quantitatively.

Mathematically proficient students make sense of quantities and their relationships in problem situations. They bring two complementary abilities to bear on problems involving quantitative relationships: the ability to *decontextualize*—to abstract a given situation and represent it symbolically and manipulate the representing symbols as if they have a life of their own, without necessarily attending to their referents—and the ability to *contextualize*, to pause as needed during the manipulation process in order to probe into the referents for the symbols involved. Quantitative reasoning entails habits of creating a coherent representation of the problem at hand; considering the units involved; attending to the meaning of quantities, not just how to compute them; and knowing and flexibly using different properties of operations and objects.

3 Construct viable arguments and critique the reasoning of others.

Mathematically proficient students understand and use stated assumptions, definitions, and previously established results in constructing arguments. They make conjectures and build a logical progression of statements to explore the truth of their conjectures. They are able to analyze situations by breaking them into cases, and can recognize and use counterexamples. They justify their conclusions,

Copyright 2010. National Governors Association Center for Best Practices and Council of Chief State School Officers. All rights reserved.

communicate them to others, and respond to the arguments of others. They reason inductively about data, making plausible arguments that take into account the context from which the data arose. Mathematically proficient students are also able to compare the effectiveness of two plausible arguments, distinguish correct logic or reasoning from that which is flawed, and—if there is a flaw in an argument—explain what it is. Elementary students can construct arguments using concrete referents such as objects, drawings, diagrams, and actions. Such arguments can make sense and be correct, even though they are not generalized or made formal until later grades. Later, students learn to determine domains to which an argument applies. Students at all grades can listen or read the arguments of others, decide whether they make sense, and ask useful questions to clarify or improve the arguments.

4 Model with mathematics.

Mathematically proficient students can apply the mathematics they know to solve problems arising in everyday life, society, and the workplace. In early grades, this might be as simple as writing an addition equation to describe a situation. In middle grades, a student might apply proportional reasoning to plan a school event or analyze a problem in the community. By high school, a student might use geometry to solve a design problem or use a function to describe how one quantity of interest depends on another. Mathematically proficient students who can apply what they know are comfortable making assumptions and approximations to simplify a complicated situation, realizing that these may need revision later. They are able to identify important quantities in a practical situation and map their relationships using such tools as diagrams, two-way tables, graphs, flowcharts and formulas. They can analyze those relationships mathematically to draw conclusions. They routinely interpret their mathematical results in the context of the situation and reflect on whether the results make sense, possibly improving the model if it has not served its purpose.

5 Use appropriate tools strategically.

Mathematically proficient students consider the available tools when solving a mathematical problem. These tools might include pencil and paper, concrete models, a ruler, a protractor, a calculator, a spreadsheet, a computer algebra system, a statistical package, or dynamic geometry software. Proficient students are sufficiently familiar with tools appropriate for their grade or course to make sound decisions about when each of these tools might be helpful, recognizing both the insight to be gained and their limitations. For example, mathematically proficient high school students analyze graphs of functions and solutions generated using a graphing calculator. They detect possible errors by strategically using estimation and other mathematical knowledge. When making mathematical models, they know that technology can enable them to visualize the results of varying assumptions, explore consequences, and compare predictions with data. Mathematically proficient students at various grade levels are able to identify relevant external mathematical resources, such as digital content located on a website, and use them to pose or solve problems. They are able to use technological tools to explore and deepen their understanding of concepts.

6 Attend to precision.

Mathematically proficient students try to communicate precisely to others. They try to use clear definitions in discussion with others and in their own reasoning. They state the meaning of the symbols they choose, including using the equal sign consistently and appropriately. They are careful about specifying units of measure, and labeling axes to clarify the correspondence with quantities in a problem. They calculate accurately and efficiently, express numerical answers with a degree of precision appropriate for the problem context. In the elementary grades, students give carefully formulated explanations to each other. By the time they reach high school they have learned to examine claims and make explicit use of definitions.

Copyright 2010. National Governors Association Center for Best Practices and Council of Chief State School Officers. All rights reserved.

7 Look for and make use of structure.
Mathematically proficient students look closely to discern a pattern or structure. Young students, for example, might notice that three and seven more is the same amount as seven and three more, or they may sort a collection of shapes according to how many sides the shapes have. Later, students will see 7×8 equals the well remembered $7 \times 5 + 7 \times 3$, in preparation for learning about the distributive property. In the expression $x^2 + 9x + 14$, older students can see the 14 as 2×7 and the 9 as $2 + 7$. They recognize the significance of an existing line in a geometric figure and can use the strategy of drawing an auxiliary line for solving problems. They also can step back for an overview and shift perspective. They can see complicated things, such as some algebraic expressions, as single objects or as being composed of several objects. For example, they can see $5 - 3(x - y)^2$ as 5 minus a positive number times a square and use that to realize that its value cannot be more than 5 for any real numbers x and y.

8 Look for and express regularity in repeated reasoning.
Mathematically proficient students notice if calculations are repeated, and look both for general methods and for shortcuts. Upper elementary students might notice when dividing 25 by 11 that they are repeating the same calculations over and over again, and conclude they have a repeating decimal. By paying attention to the calculation of slope as they repeatedly check whether points are on the line through (1, 2) with slope 3, middle school students might abstract the equation $(y - 2)/(x - 1) = 3$. Noticing the regularity in the way terms cancel when expanding $(x - 1)(x + 1)$, $(x - 1)(x^2 + x + 1)$, and $(x - 1)(x^3 + x^2 + x + 1)$ might lead them to the general formula for the sum of a geometric series. As they work to solve a problem, mathematically proficient students maintain oversight of the process, while attending to the details. They continually evaluate the reasonableness of their intermediate results.

Connecting the Standards for Mathematical Practice to the Standards for Mathematical Content
The Standards for Mathematical Practice describe ways in which developing student practitioners of the discipline of mathematics increasingly ought to engage with the subject matter as they grow in mathematical maturity and expertise throughout the elementary, middle and high school years. Designers of curricula, assessments, and professional development should all attend to the need to connect the mathematical practices to mathematical content in mathematics instruction.

The Standards for Mathematical Content are a balanced combination of procedure and understanding. Expectations that begin with the word "understand" are often especially good opportunities to connect the practices to the content. Students who lack understanding of a topic may rely on procedures too heavily. Without a flexible base from which to work, they may be less likely to consider analogous problems, represent problems coherently, justify conclusions, apply the mathematics to practical situations, use technology mindfully to work with the mathematics, explain the mathematics accurately to other students, step back for an overview, or deviate from a known procedure to find a shortcut. In short, a lack of understanding effectively prevents a student from engaging in the mathematical practices.

In this respect, those content standards which set an expectation of understanding are potential "points of intersection" between the Standards for Mathematical Content and the Standards for Mathematical Practice. These points of intersection are intended to be weighted toward central and generative concepts in the school mathematics curriculum that most merit the time, resources, innovative energies, and focus necessary to qualitatively improve the curriculum, instruction, assessment, professional development, and student achievement in mathematics.

Copyright 2010. National Governors Association Center for Best Practices and Council of Chief State School Officers. All rights reserved.

Chapter One : Infinity

Class Activity 1:   The Haunted House

Just go on and faith will soon return.

Jean D'Alembert

It is a dark and stormy night. You are standing before a creepy old mansion with vandalism in your heart. The windows are shadowed, and vines tangle over the stone façade. For some reason you have an infinite pile of stones at your feet and a permanent marker in your hand. You pick up a stone, write the number "1" on it, then you pick up another, write "2" on it and you pitch both stones through a jagged window. Suddenly, stone number 1 is returned to you, thrown back out through the window by a spirit unseen. So, you pick up a couple more stones, write "3" and "4" on them and hurl them through the window, satisfied by the tinkle of shattered glass. Stone "2" is returned to you. Not to be outdone, you grab "5" and "6" and send them in. Out comes "3." If you and the spirit continue this game *forever*, how many stones end up in the house?

Let's practice those debating skills:

a) Try to argue the answer is "infinitely many."
b) Now try to argue that the answer is zero.
c) Try to argue something in between.
d) Try to argue that the problem makes no sense.

Read and Study

It takes me maybe an hour to understand a page of new mathematics. At the end, if I'm lucky I obviously say to myself "Why didn't I understand that right away? What took the hour?"

Marvin Minsky

The idea of infinity brings with it lots of strange possibilities. Consider the Haunted House activity. Perhaps in the end there are infinitely many stones in the house – after all, a stone accumulates there with each round of tosses. On the other hand, if you try to name a *specific* stone that ends up in the house, you'll find you cannot. Number 5600 is not in there; it comes out on the 5600th exchange of tosses. Neither is stone number 1,000,000 or number 1,000,000,000.

Historically, some mathematicians have argued that problems like the Haunted House activity make no sense because an infinite process *cannot end*. Consider the words of one of the most famous mathematicians of all time, Carl Friedrich Gauss (1777 – 1855):

> …I protest above all against the use of an infinite quantity as a *completed* one, which in mathematics is never allowed. The infinite is only a manner of speaking ….

Since Gauss' time, mathematicians have found ways to deal with an infinite process as a *completed* one, but if you don't like the idea at this point, you are in good company. Gauss is so famous that he was even on the German 10 mark note (before they switched to euro notes in 2002.

Zeno, an early Greek thinker, is remembered for describing more paradoxes involving infinity. His paradoxes deal with motion and are rooted in hard questions about the nature of time and space. Three of his most famous include the following:

- **Dichotomy:** It is impossible to cover any distance, because half the distance must be traversed first, then half the remaining distance, then again half of what remains, and so on. Some portion of the distance to be covered is *always* left to cover. Therefore, motion is impossible.

- **Achilles and the Tortoise:** Achilles runs to overtake the tortoise, but he must reach the point where the tortoise started, from which it has already departed. Repeating indefinitely, Achilles gets to each new point in the race, the tortoise having been there has already left. Therefore, even though Achilles is much faster than the tortoise, he can never overtake the tortoise.

- **The Arrow:** An arrow shot from a bow must be moving at every instant in its flight. But at every instant it must be somewhere in space. However, if it is always in some specific place, it can't be in transit at every instant, for to be in transit is to be nowhere. In other words, in any instant of time there is no motion occurring. If the arrow cannot move in a single instant, then it cannot move in any instant, and thus the arrow cannot move.

(A paradox is a proposition that seems self-contradictory or absurd, but expresses a possible truth. *Think about each of Zeno's paradoxes above and make sure you understand both the absurdity and the possible truth.*)

The point of all the above discussion is to acknowledge that thinking about infinity is hard. In this book, we will develop some tools to make it easier, and we will show you that it is possible to solve lots of neat problems involving motion. In fact, that is what calculus is all about.

Each of Zeno's paradoxes views motion as an infinite process. We use the term *infinite process* to describe a procedure that continues on forever. A simple example of an infinite process would be to start with the number one, then continue to add one forever. We could write this as

$$'1 + 1 + 1 + ...'$$

The "dot dot dot" conveys the idea that we keep adding ones *indefinitely* (forever).

First, we want to say up front that **infinity is not a number;** it is an *idea*. (A big idea!) For example, a mathematician would *not* say that the sum $1 + 2 + 3 + 4 + 5 + ...$ *equals* infinity. We might say that the sum diverges or becomes arbitrarily large. More precisely, this means that for all numbers, n, the sum is eventually bigger than n. This kind of thinking and language is helpful; use it.

The reason why it may seem to you that infinity is a number is because we often say there is an infinite number of something. For example, there are an infinite number of prime numbers, and there are an infinite number of points on a line. Another way to say this is that there are infinitely many prime numbers and infinitely many points on a line. More precisely, we are talking about the cardinality of sets.

The **cardinality** of a set is the number of elements in a set. A **finite set** is a set with exactly n elements, for some whole number n. For example, the set of vertices on a cube is finite, with cardinality 8. *Create a set with a **cardinality** of 7.*

While we will study cardinality and infinite sets in more depth later in the chapter, for now we can say that a set is **infinite** if for all numbers, n, the cardinality of the set is bigger than n.

Before we go any further, you should know that reading a math book takes time. You do not read it like an ordinary book. You read it slowly – really slowly – with a pencil in hand. You write on it. Forget what you were taught about not writing in books – mathematicians write all over their texts. We write in the margins. We underline stuff. We are going to help you learn to read this way first, by making the reading sections short, and second, by asking you lots of italicized questions that you should stop and answer as you read. Some of the questions might take you twenty minutes or more to think through. Please take the time to do it.

Homework

The greatest mistake you can make in life is continually fearing that you will make one.
Ellen Hubbard

1) Do all the italicized things in the *Read and Study* section.

2) Decide whether each of the following statements is True or Flase. In each case, explain your reasoning
   a) 'Infinity' is the biggest number.
   b) π is an example of an infinite number.

3) Suppose that Achilles can run 10 meters per second, the tortoise only 5 meters per second, and that the track is 100 meters long. Furthermore, Achilles, being the fair sportsman that he is, gives the tortoise a 10-meter advantage. According to Zeno, the tortoise wins the race! Achilles cannot believe this result. Who *will* win? Why? Where is the "mistake" in Zeno's reasoning? Can you identify the infinite process in this problem?

4) A **prime number** is a positive **integer** that has exactly two distinct **factors**. (When we bold and italicize a term we are showing you that there is a precise mathematical definition for that term that you should know well enough to write and explain to a classmate. These terms can be found in the glossary. *Look up and learn the three bolded terms now.*

5) The Greek mathematician, Euclid, proved (around 300 BC) that there are infinitely many of these prime numbers. Using the language described in the *Read and Study* section, what precisely does it mean to say that there are infinitely many primes?

6) One hundred years after Euclid, Eratosthenes, another Greek mathematician, developed an ideal method for listing all the prime numbers less than a given integer *n*. His method is called the **Sieve of Eratosthenes**. He suggested that to find prime numbers, one should begin with a list of all the positive integers (except 1). Find the first number on the list, circle it, and then cross-out all multiples of that number. Then do the same with the next remaining number on the list, and then the next, and the next ... In the end, only the primes will remain.

   a) Try it below to find all primes less than 200. Can you see why he called this a 'sieve?' (The sieve appears in the *Introduction to Arithmetic* by Nicomedes.)

1	2	3	4	5	6	7	8	9	10
11	12	13	14	15	16	17	18	19	20
21	22	23	24	25	26	27	28	29	30
31	32	33	34	35	36	37	38	39	40
41	42	43	44	45	46	47	48	49	50
51	52	53	54	55	56	57	58	59	60
61	62	63	64	65	66	67	68	69	70
71	72	73	74	75	76	77	78	79	80
81	82	83	84	85	86	87	88	89	90
91	92	93	94	95	96	97	98	99	100
101	102	103	104	105	106	107	108	109	110
111	112	113	114	115	116	117	118	119	120
121	122	123	124	125	126	127	128	129	130
131	132	133	134	135	136	137	138	139	140
141	142	143	144	145	146	147	148	149	150
151	152	153	154	155	156	157	158	159	160
161	162	163	164	165	166	167	168	169	170
171	172	173	174	175	176	177	178	179	180
181	182	183	184	185	186	187	188	189	190
191	192	193	194	195	196	197	198	199	200

   b) Why does the method work to reveal the primes less than *n*?

   c) Since this method works for every integer *n*, does it constitute a proof that there are infinitely many primes? Explain your thinking.

Class Activity 2: So Many Primes!

3 is a prime, 5 is a prime, 7 is a prime. Why bother with non-prime numbers when the primes can do everything?

William of Ockham

Euclid's proof that there are infinitely many primes depends on the **Fundamental Theorem of Arithmetic**. This theorem says that every positive integer greater than 1 can be written in exactly one way as a product of primes.

Just for practice, write 480 as a product of primes.

Now back to Euclid. He began his proof by assuming that there *is* a largest prime number. He then showed that this assumption together with the Fundamental Theorem of Arithmetic led to a contradiction. Therefore his original assumption must be false. (This method of proof is called **proof by contradiction**.) Here is the proof. It is your group's job to study it and answer the italicized questions.

> **Theorem:** There are infinitely many prime numbers.
>
> **Proof:** Suppose not. Suppose that there are only finitely many primes, and so there is some largest prime number, which we will call P. Then the following list:
> $$2, 3, 5, 7, 11, \ldots, P$$
> is the *complete list* of all of the primes.
>
> Now consider the number that is the product of all of these primes plus one; we'll call it Q.
>
> $$Q = [2 \times 3 \times 5 \times 7 \times 11 \times 13 \times 17 \ldots \times P] + 1.$$
>
> Now Q is surely bigger than P and so Q cannot be prime (*Why not?*).
>
> So Q must be composite, and that means it can be written as the product of primes. (*Why?*)
>
> In other words, there is at least one prime that is a factor of Q. Which one is it?

Is it possible that 2 is a factor of Q? (*Have a look at the form of Q to decide and explain why.*)

Is it possible that 3 is a factor of Q? (*Have a look at the form of Q to decide and explain why.*) *Make sure everyone understands this argument.*

We can make similar arguments to show that **none** of the numbers on our list is of primes 2, 3, 5, 7, 11, ..., P can be a factor of Q.

But since Q is not prime (*how do we know Q isn't prime?*), there must be another prime larger than P that is a factor of Q. *Why is this?*

This is a contradiction. *Identify precisely what has been contradicted.*

So our first assumption that there are only finitely many primes must be wrong.

So there must be infinitely many primes.

Read and Study

"Contrariwise,' continued Tweedledee, 'if it was so, it might be; and if it were so, it would be; but as it isn't, it ain't. That's logic."
  Lewis Carroll, Alice's Adventures in Wonderland & Through the Looking-Glass

The process of doing mathematics has a language of its own. We'll remind you of the important terms in this section. *First, study these definitions and make sure you understand them.*

1) **axiom:** A mathematical statement or hypothesis we *assume* is true. (In other words, we agree to accept it without requiring a proof.) For example, it is an axiom of Euclidean Geometry that a unique line can be drawn between any two points.

2) **inductive reasoning:** coming to a conclusion based on examples. For instance, I observe that 3, 5 and 7 are all prime. Now, based on these examples I might reason (incorrectly, by the way) that all odd numbers are prime. Or I might notice that the sun rose day before yesterday, it rose yesterday, it rose today. So I might conclude that the sun will rise tomorrow. This is inductive reasoning. This type of reasoning has a place in mathematics – we use it in making conjectures. But we do not accept it as a suitable proof unless we can examine all the possible examples.

3) **deductive reasoning:** coming to a conclusion based on logic. Here is a trivial example: Madison is in Wisconsin. Wisconsin is in the United States. Therefore, Madison is in the United States. For a more interesting example, see the argument in 7) below.

4) **conjecture:** a conjecture is an hypothesis or a guess about what is true. For example, after some experience with multiplication, a student might *conjecture* that multiplying two numbers always results in a product that is bigger than either of the original numbers. Conjectures are often made based on inductive reasoning.

5) **counterexample:** a counterexample is an example that shows a conjecture is false. For example, $(1/2) \times (6) = 3$ and so the above conjecture is shown to be false.

6) **proof:** a mathematical proof consists of a deductive argument based on definitions, axioms and theorems that establishes the truth of a claim.

7) **theorem:** a theorem is a mathematical statement that has been proven true. For example, it is a theorem that the product of two odd numbers is always an odd number. (Here is a proof: suppose you have two odd numbers, one of the form $2a + 1$ and one of the form $2b + 1$ where a and b are both whole numbers. (*Why can we assume all odd numbers have this form?*) So $(2a + 1) \times (2b + 1) = 4ab + 2a + 2b + 1 = 2(2ab + a + b) + 1$. And this also has the *form* of an odd number. Note that this is an example of deductive reasoning.)

The use of good mathematical language and notation are *habits*. You cannot speak and write messy mathematics in your college courses and then magically expect to use good language and notation in front of your class. You must practice.

Many mathematical statements are conditional statements ("if-then" statements). Here are some examples. *Decide whether each is true or false, and in each case explain your thinking.*

1) If a polygon is a square, then it is a rectangle.

2) If a polygon is a rectangle, then it is a square.

3) If you live in Los Angeles, then you live in California.

4) If you don't live in California, then you don't live in Los Angeles.

5) If it is raining, then I have my umbrella.

Mathematicians call the statement 'If Q, then P' the **converse** of the statement 'If P, then Q'. So Statement 2 is the converse of Statement 1 above. Note that those two statements are not **logically equivalent** (not always true at the same time.).

A statement of the form 'If not Q, then not P' is called the **contrapositive** of the statement 'If P, then Q.' The contrapositive form *is* logically equivalent to 'If P, then Q". So Statement 3 and Statement 4 above are logically equivalent statements. *Think about it to make sure that seems right. Write the converse of Statement 5 above.*

Now write the contrapositive of Statement 5 above.

Now stop and think. Do you see that the converse means something different from the original statement? Explain why.

Sometimes when you want to prove that an if/then statement is true, it is easier to prove that the contrapositive statement is true than it is to prove the original statement is true. And the point here is that by proving the contrapositive you also prove the original (because they are equivalent statements).

Connections to Teaching

Those who know how to think need no teachers.
                                                            Mahatma Gandhi

The National Council of Teachers of Mathematics (NCTM) in their *Principles and Standards for School Mathematics* (2000) asserts that students in grades six through eight should:

"Recognize reasoning and proof as fundamental aspects of doing mathematics."

"Make and investigate mathematical conjectures."

"Develop and evaluate mathematical arguments and proofs."

"Select and use various types of reasoning and methods of proof." (p. 262)

In other words, NCTM advocates that middle grades students should learn to *think* like mathematicians. Teaching your students to think like this will be part of your job. They will learn this by engaging in activities designed to introduce them to the language and ideas of mathematics, and they will learn by mimicking the ways that you think and talk about mathematics. You will be their mathematical role model. They will be watching you – and learning about how mathematical problems are posed, how problems are attacked, what words are used to describe the process of doing mathematics, whether more than one method of solution is acceptable, and how arguments are made and challenged. As a teacher, you will need to remember that you are being watched and that kids are always learning *something* - even things you are not explicitly trying to teach.

We realize that you may not always have had good mathematical role models. So let us spell a few things out:

Mathematics is *not* all about numbers. It *is* all about patterns.

Mathematicians look for patterns and *create* general formulas based on our understanding of the structure of those patterns. We do not simply memorize and apply formulas. Mathematicians value careful definitions of mathematical objects. A definition should give the precise criteria needed to classify an object and be useful in making a deductive argument.

Mathematicians care about *why* a conjecture is true and we always want a proof. We can't stress this enough.

Mathematicians value a variety of solution strategies. We often try to solve a problem or prove a theorem in more than one way. (For instance, there are more than four hundred known proofs of the Pythagorean Theorem.)

Mathematicians like to create examples of and models for mathematical ideas.

Think about each of the above assertions. Try to create examples to help you make sense of them, and be sure to discuss these in class.

Homework

Success is following the pattern of life one enjoys most.
Al Capp

1) Do all the italicized things in the *Read and Study* and *Connections to Teaching* sections.

2) Make up new examples to help teach your students the difference between inductive reasoning and deductive reasoning.

3) In your own words, explain the general idea of proof by contradiction.

4) We have a conjecture: All prime numbers are odd. Prove it or find a counterexample.

5) Write the **converse** of the conjecture in problem 4) above. Prove the converse if true or find a counterexample.

6) Write the **contrapositive** of the conjecture in problem 4) above. Prove the converse if true or find a counterexample

7) We have another conjecture: If two numbers are even, then so is their product. Prove it or find a counterexample.

8) One more conjecture! If 2 is a factor of n^2, then 2 is a factor n. Prove it or find a counterexample.

Class Activity 3: Exploring Infinite Lists

A **sequence** is an infinite ordered list of numbers $a_1, a_2, a_3, a_4, \ldots$, which are called the **terms** of the sequence. Another way to think of a sequence is as a function with a domain of the set of natural numbers. Given a natural number n, the nth term of the sequence is denoted a_n.

We say a sequence **converges** if the terms get arbitrarily close to some fixed value, and we say a sequence **diverges** if it does not converge.

Below are some examples of sequences. For each, determine a formula for the nth term a_n as a function of n. Then decide whether the sequence converge or diverges.

1) $1, \frac{1}{2}, \frac{1}{3}, \frac{1}{4}, \frac{1}{5}, \ldots$

2) $1, \sqrt{2}, \sqrt{3}, \sqrt{4}, \sqrt{5}, \ldots$

3) $\frac{3}{2}, \frac{6}{3}, \frac{9}{4}, \frac{12}{5}, \frac{15}{6}, \ldots$

4) $3, 2, 3, 2, 3, 2, 3, 2, \ldots$

5) $-1, \frac{3}{4}, -\frac{4}{8}, \frac{5}{16}, -\frac{6}{32}, \ldots$

6) $3, 2, \frac{4}{3}, \frac{8}{9}, \frac{16}{27}, \ldots$

Now let's try to make sense of a formal definition:

> **Definition:** A sequence a_n **converges** to a **limit** L means that a_n is arbitrarily close to L (within any given distance ε away) for all n sufficiently large (greater than some value N).

For each sequence above that you decided converges, use a calculator or spreadsheet to determine the value of the limit L, and then determine the value of N required so that the terms a_n will always be within $\varepsilon = \frac{1}{10}$ of L for all n greater than N.

Read and Study

A **sequence** is an infinite ordered list of numbers $a_1, a_2, a_3, a_4, \ldots$, which are called the **terms** of the sequence. Another way to think of a sequence is as a function with a domain of the set of natural numbers. Given a natural number n, the nth term of the sequence is denoted a_n.

We say a sequence **converges** if the terms get arbitrarily close to some fixed value, and we say a sequence **diverges** if it does not converge. To make the definition of convergence more precise, we will make this formal definition:

> **Definition:** A sequence a_n **converges** to a **limit** L means that a_n is arbitrarily close to L (within any given distance ε away) for all n sufficiently large (greater than some value N).

In other words, a sequence a_n **converges** to a **limit** L means that if were to pick how close you want the terms to be to L (that's what the ε means), you can find how far along the sequence you need to go (that's what the N means) until all the remaining terms will be within that distance of the limit.

The symbol ε is pronounced "epsilon", and it's the Greek letter that corresponds to our Latin letter "e". In mathematics, ε (epsilon) is traditionally used to denote a small number that could be chosen to be arbitrarily close to zero.

To illustrate what this all means, let's consider the following sequence: $1, \frac{1}{2}, \frac{1}{3}, \frac{1}{4}, \frac{1}{5}, \frac{1}{6}, \ldots$

A formula for the nth term of this sequence is $a_n = \frac{1}{n}$. As n grows larger and larger (tends to infinity), the terms of this sequence get smaller and smaller, and tend to 0. Here is a graph of the first 100 terms of the sequence, with the values of a_n on the vertical axis and n on the horizontal.

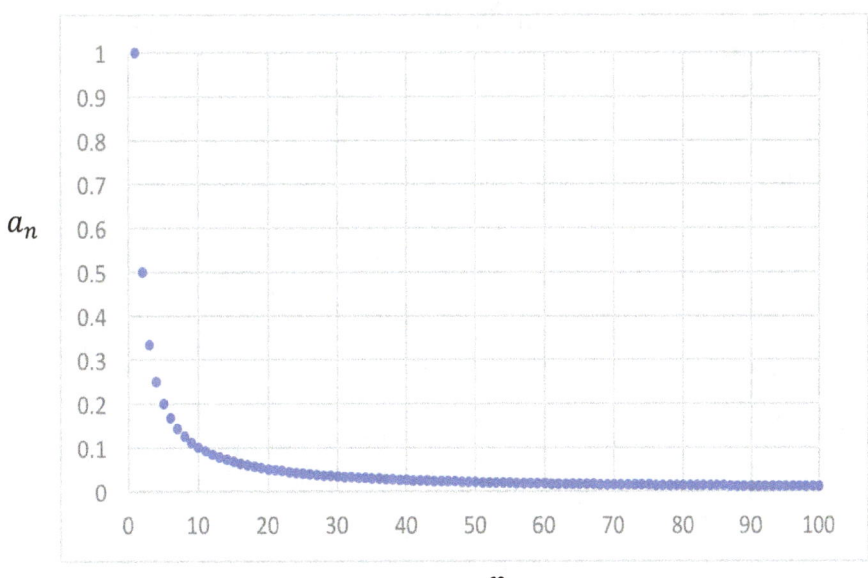

Suppose we wanted to the sequence to be within $\varepsilon = 0.10$ of the limit $L = 0$. How far down the sequence do we need to go before the sequence is within that distance of the limit? Well, when $n = 10$, the sequence value is $1/10$, which is exactly the distance of $\varepsilon = 0.10$ away from $L = 0$, and when $n > 10$, the sequence value will be less than $1/10$, and so within $\varepsilon = 0.10$ away from $L = 0$. So if we want to be within $\varepsilon = 0.10 = 1/10$ of the limit, we just need to go beyond the N = 10th term of the sequence. Here's a graph of just the first 30 terms, with a red line at the limiting value limit L = 0 and a green line at the distance of $\varepsilon = 0.10$ away from the limit.

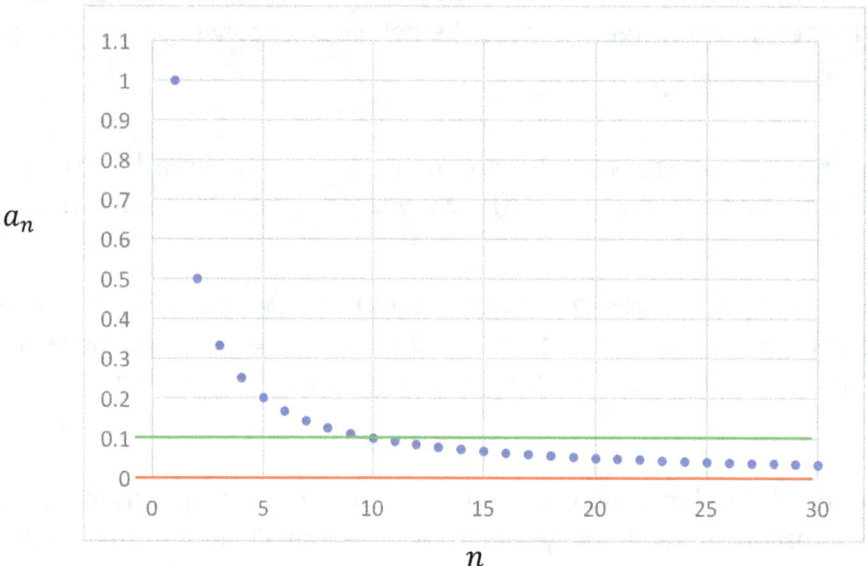

Now suppose wanted to the sequence to be within $\varepsilon = 0.02$ of the limit $L = 0$. How far down the sequence do we need to go before the sequence is within this distance from the limit? Well, when $n = 50$, the sequence value is $1/50$ which is exactly 0.02 away from the limit of zero, and when $n > 50$, the sequence value will be less than $1/50$, and so within $\varepsilon = 0.02$ from the limit $L = 0$. So if we want to be within $\varepsilon = 0.02 = 1/50$ of the limit, we just need to go beyond the N = 50th term of the sequence. Here's a close-up of the graph of the terms of the sequence $a_n = \frac{1}{n}$ from $n = 10$ to $n = 100$, with a red line at the limiting value limit L = 0 and a green line at the distance of $\varepsilon = 0.02$ away from the limit.

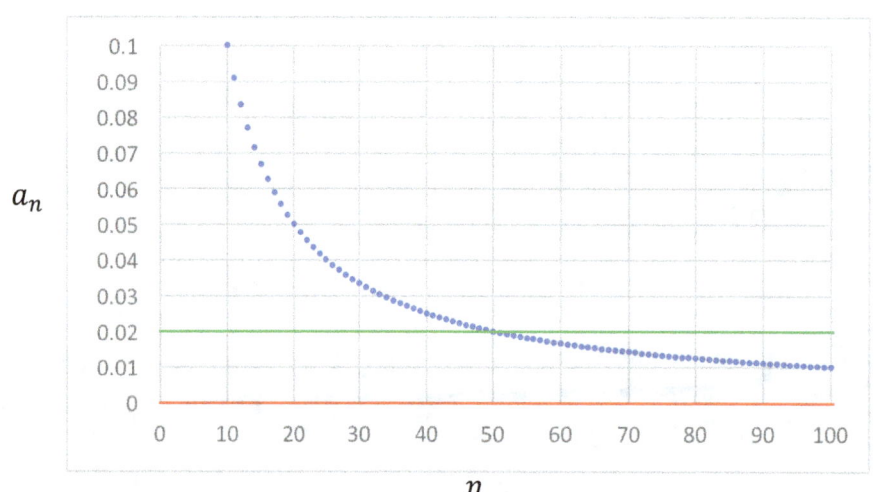

In general, whatever ε is given, no matter how small ε may be, for this sequence we know that can just choose $N=1/\varepsilon$ and then we know the terms will be within ε of $L = 0$ for all $n > N$. That's what is means to prove that the sequence $a_n = \frac{1}{n}$ converges.

Let's look at a new example. Consider the following sequence:

$$\frac{5}{5}, \frac{10}{7}, \frac{15}{9}, \frac{20}{11}, \frac{25}{13}, \ldots$$

A formula for the nth term would be $a_n = \frac{5n}{2n+3}$. Check to see that this formula does indeed generate the terms shown above. We put this formula into a spreadsheet and generated a graph of the first 50 terms of the sequence, which is shown below:

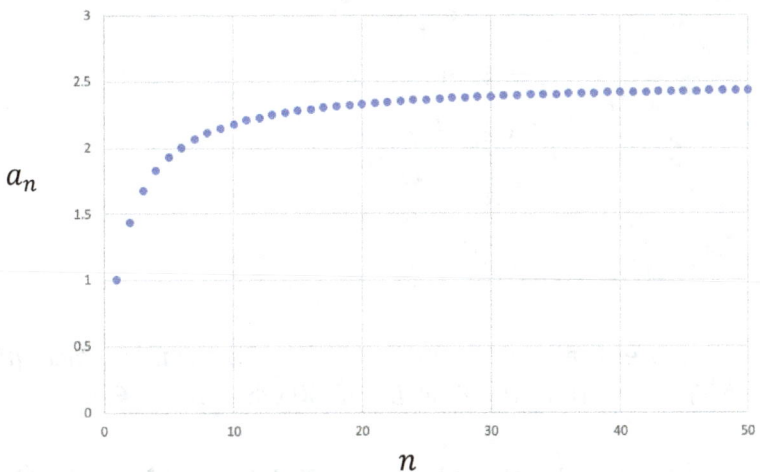

From the graph, it rather looks like the sequence is converging with a limit of exactly 2.50 or $\frac{5}{2}$. How do we know it won't continue to grow beyond $\frac{5}{2}$? We'll let you think about that. *Make an argument that the expression $\frac{5n}{2n+3}$ will always be a number less than $\frac{5}{2}$.*

Now let's illustrate the definition of convergence by finding exactly how far we need to go along the sequence to get within a particular epilson of this limit. Suppose first we want to be within $\varepsilon = 0.10$ of the limit $L = 2.50$. Well, when $n = 36$, the sequence value is 2.40 (*check to make sure that is right*), which is exactly the distance of $\varepsilon = 0.10$ away from $L = 2.50$. Here's that graph of just the first 50 terms, with a red line at the limiting value limit L = 2.50 and green lines at the distance of $\varepsilon = 0.10$ away from that limit.

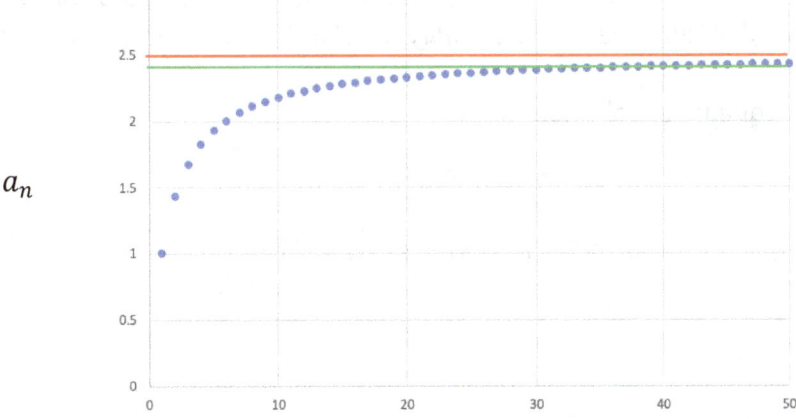

25

Now suppose we want to be within $\varepsilon = 0.01$ of the limit $L = 2.50$. Doing some more calculations on our spreadsheet, we find that when $n = 373$, the sequence value is approximately 2.489986 and when $n = 374$, the sequence value is approximately 2.490013. So it looks like the sequence gets within a distance of $\varepsilon = 0.01$ away from $L = 2.50$ when $n \geq 374$.

Note: we could have found the value $n = 374$ without using the spreadheet by instead doing some algebra ourselves. We want to find the value for n such that $\frac{5n}{2n+3}$ first gets within 0.01 of 2.50, in other words, we could just solve the equation $\frac{5n}{2n+3} = 2.49$ as follows:

$$\frac{5n}{2n+3} = 2.49$$

$$5n = 2.49(2n+3)$$

$$5n = 4.98n + 7.47$$

$$0.02n = 7.47$$

$$n = \frac{7.47}{0.02}$$

$$n = 373.5$$

So then the sequence will be greater than 2.49 when $n \geq 374$. *Do a similar calculation to confirm that the sequence is first within $\varepsilon = 0.10$ of the limit $L = 2.50$ when $n = 36$.*

So we've now gotten really quite close to 2.50, but what if someone wants to be within $\varepsilon = 0.0000001$ or $\varepsilon = 0.0000000000001$ of this limit? A way we can make the argument to prove that the limit is indeed $\frac{5}{2}$ is to do a little algebra, like we did above for $\varepsilon = 0.01$, and you did above for $\varepsilon = 0.10$, but now for any ε. We'll just keep the ε unspecified in our calculations. To prove that limit of the sequence with terms $a_n = \frac{5n}{2n+3}$ is $L = \frac{5}{2}$, means for any ε we are given, we need to find the value of n so that $\frac{5n}{2n+3}$ will always be within ε of the limit $L = \frac{5}{2}$. Since we know $\frac{5n}{2n+3}$ is always less than 5/2 (you made that argument earlier, right?), we need to find n so that

$$\frac{5n}{2n+3} > \frac{5}{2} - \varepsilon$$

Make sure you understand why that is the right inequality to describe our situation. Why are we subtracting epsilon from 5/2? Why does the inequality point in that direction and not the other?

Now we can solve the inequality as follows:

$$\frac{5n}{2n+3} > \frac{5}{2} - \varepsilon$$

$$5n > \frac{5}{2}(2n+3) - \varepsilon(2n+3)$$

$$5n > 5n + \frac{15}{2} - 2\varepsilon n - 3\varepsilon$$

$$2\varepsilon n > \frac{15}{2} - 3\varepsilon$$

$$n > \frac{15}{4\varepsilon} - \frac{3}{2}$$

$$n > \frac{15-6\varepsilon}{4\varepsilon}.$$

Our algebra above shows that $\frac{5n}{2n+3}$ will be within ε of $\frac{5}{2}$ whenever $n > \frac{15-6\varepsilon}{4\varepsilon}$.

Read again the formal definition of convergence (below) and thoroughly check that the preceding sentence we wrote above is exactly what proves that our sequence converges to $\frac{5}{2}$.

> **Definition:** A sequence a_n **converges** to a **limit** L means that a_n is arbitrarily close to L (within any given distance ε away) for all n sufficiently large (greater than some value N).

Homework

Most of our obstacles would melt away if, instead of cowering before them, we should make up our minds to walk boldly through them.
Orison Swett Marden

1. Do all the italicized in the *Read and Study* section.

2. For each sequence, do the following:
   a) List the exact values for the first three terms of the sequence. (Do this mentally or by hand, not with a computer).
   b) Use Excel or Google Sheets to make a graph of the sequence. Show enough terms to convey the general behavior of the sequence.
   c) Based on the graph, does the sequence converge to one fixed value L? If so what is the value of the limit L? Then:
      o Using $\varepsilon = 0.1$, determine the value N so that a_n is within the distance ε away from L for all n greater than N.
      o Using $\varepsilon = 0.001$, determine the value N so that a_n is within the distance ε away from L for all n greater than N.
   d) If the sequence does not converge, explain how you can tell.

a. $a_n = \dfrac{144}{n^2}$

b. $a_n = \left(\dfrac{11}{10}\right)^n$

c. $a_n = \dfrac{n+(-1)^n}{n}$

3. For the sequence in 2a, finding an algebraic formula for the value of N as a function of ε. Write a sentence that explains how your formula proves your limit is correct.

4. For each sequence, do the following:
   e) Write a formula for the nth term of the sequence a_n.
   f) Use Excel or Google Sheets to make a graph of the sequence. Show enough terms to convey the general behavior of the sequence.
   g) Based on the graph, does the sequence converge to one fixed value L? If so what is the value of the limit L? Then:
      o Using $\varepsilon = 0.1$, determine the value N so that a_n is within the distance ε away from L for all n greater than N.
      o Using $\varepsilon = 0.001$, determine the value N so that a_n is within the distance ε away from L for all n greater than N.
   h) If the sequence does not converge, explain how you can tell.

   a. $\dfrac{1}{1}, \dfrac{2}{3}, \dfrac{3}{5}, \dfrac{4}{7}, \dfrac{5}{9}, \ldots$

   b. $\dfrac{9}{10}, \dfrac{81}{100}, \dfrac{729}{1000}, \dfrac{6561}{10000}, \ldots$

   c. $0, \dfrac{1}{2}, -\dfrac{2}{3}, \dfrac{3}{4}, -\dfrac{4}{5}, \dfrac{5}{6}, -\dfrac{6}{7}, \ldots$

5. For the sequence in 4a, finding an algebraic formula for the value of N as a function of ε. Write a sentence that explains how your formula proves your limit is correct.

6. Decide whether each of the following statements is True or False. In each case, explain your answer.

   a. If a sequence does not converge, then it must diverge.
   b. If the terms of a sequence continue to grow larger and larger, then the sequence must diverge.
   c. If a sequence diverges, then the terms must aways continue to grow larger and larger.

Class Activity 4:   Exploring Infinite Sums

In most sciences one generation tears down what another has built, and what one has established another undoes. In mathematics alone each generation adds a new story to the old structure.

Hermann Hankel

Is it possible to add up infinitely many positive numbers and get a finite sum? Discuss this in your group before you read further.

A **series** is an infinite number of ordered terms added together. The term *infinite series* is often used to emphasize the fact that series contain an infinite number of terms. Get out your calculator and use it to see whether any of these might add to a finite sum.

1) $\frac{1}{2} + \frac{1}{4} + \frac{1}{8} + \frac{1}{16} + \ldots$

2) $1 + \frac{1}{2} + \frac{1}{3} + \frac{1}{4} + \frac{1}{5} + \cdots$

3) $1 + 2 + 3 + 4 + 5 + 6 + \cdots$

4) $\frac{1}{2} + \frac{1}{2} + \frac{1}{2} + \frac{1}{2} + \cdots$

5) $.9 + .09 + .009 + .0009 + .00009 + \cdots$

6) $\frac{3}{4} + \frac{5}{8} + \frac{9}{16} + \frac{17}{32} + \frac{33}{64} + \cdots$

After your group has made a conjecture about each series, decide whether each of the following is true or false. In each case, be sure to explain your thinking.

- True or False? If the **terms** of a series eventually get arbitrarily close to zero, then the series must have a finite sum.

- True or False? If a series has a finite sum, then the terms must get arbitrarily close to zero.

Read and Study

"Can you do addition?" the White Queen asked. "What's one and one and one and one and one and one and one and one and one and one?"
"I don't know," said Alice. "I lost count."

Lewis Carroll
Through the Looking Glass

Is it possible to add up infinitely many positive numbers and end up with a finite sum? The answer is 'yes,' and, in that case, we say that the series converges.

In the previous section, we made a formal definition for what it means for an infinite *sequence* to converge. Well now we will say an infinite *series* is said to converge if its **sequence of partial sums** converges. An infinite series that does not converge is said to diverge.

What we mean by "arbitrarily close" is this: we can make the partial sum as close as we want to the exact sum by adding up enough terms in the series. For example, the first series in the class activity, $\frac{1}{2}+\frac{1}{4}+\frac{1}{8}+\frac{1}{16}+\ldots$ converges to 1 because the sequence of partial sums gets arbitrary close to 1.

Let's have a look. The first four partial sums are given below.

$$\frac{1}{2} = \frac{1}{2}$$
$$\frac{1}{2}+\frac{1}{4} = \frac{3}{4}$$
$$\frac{1}{2}+\frac{1}{4}+\frac{1}{8} = \frac{7}{8}$$
$$\frac{1}{2}+\frac{1}{4}+\frac{1}{8}+\frac{1}{16} = \frac{15}{16}$$
$$\vdots = \vdots$$
$$\frac{1}{2}+\frac{1}{4}+\frac{1}{8}+\frac{1}{16}+\cdots = 1$$

We claimed that the partial sums will get arbitrarily close to 1. *Think about this. Suppose we wanted the partial sum within 0.01 of 1. How many terms would we need at a minimum to do this?*

Now let's look at a geometric proof that Series 1 converges. *Take a look at the picture below and try to understand it.*

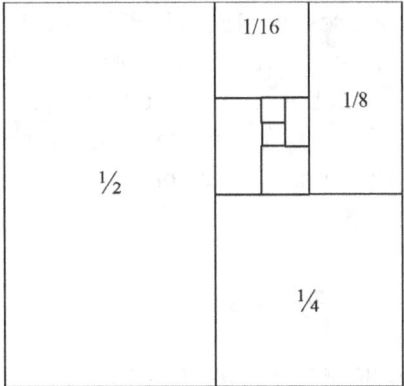

You will need to get used to the idea that when a series gets "arbitrarily close" to a number, even though the partial sum may never quite equal that number, we still say that the number is the *exact sum* of the infinite series.

So mathematicians would say that the sum of the series in the above example *equals* 1, and write

$$\frac{1}{2}+\frac{1}{4}+\frac{1}{8}+\frac{1}{16}+ \ldots = 1$$

Now let's consider the series 0.9 + 0.09 + 0.009 + 0.0009 … The sequence of partial sums gets arbitrarily close to 1, and so we say the sum of this series is 1, and write 0.999… = 1.

Let us say this again: mathematicians do not say 0.999… is a little less than 1. We say it is *exactly* 1. Think of it like this: Two numbers are **equal** if there is no positive distance between them on the number line. So the question becomes: how far apart are 0.999… and 1? It turns out that for every positive distance you can name, 0.999 … and 1 are *closer than that distance*.

You have seen some series that converge and some series that diverge, and you might ask, is there a quick and easy way to tell if a given series will converge?

The first thing that we note is that there is no chance that the series will converge unless the terms are getting smaller and smaller and smaller… so Series 3 and Series 4 from the start of this section are not going to converge. But having the terms decrease is not enough. *Look at Series 6.* Here the terms are decreasing by getting closer and closer to ½. In order for a series to have a prayer at having a finite sum, the terms need to get closer and closer to *zero*. Only the Series 1, 2, and 5 meet this criterion.

But here is the frustrating thing: even if the terms go to zero, this is *not* enough to ensure convergence. You see, while Series 1 and 5 both converge, we will now prove that Series 2 (it is called the **harmonic series**) diverges. In other words, the harmonic series grows without bound even though the terms get closer and closer to zero.

Here is a classic proof by Nicole d'Oresme (1323-1382) that the harmonic series diverges (Dunham, 1990). (Remember that we call the statement *'the harmonic series diverges'* a **theorem** because it has been proved based on logic.) *Your job is to make sure that you understand this proof well enough to explain it. Don't memorize it – that's not the point. Make sure that you understand both the details (how one line follows from the lines above it) and the big idea.*

Theorem: The harmonic series diverges.

Proof: Here is the harmonic series grouped in a special way:

$$1 + \left(\frac{1}{2}\right) + \left(\frac{1}{3} + \frac{1}{4}\right) + \left(\frac{1}{5} + \frac{1}{6} + \frac{1}{7} + \frac{1}{8}\right) + \left(\frac{1}{9} + \frac{1}{10} + \frac{1}{11} + \frac{1}{12} + \frac{1}{13} + \frac{1}{14} + \frac{1}{15} + \frac{1}{16}\right) + \cdots$$

Compare the harmonic series to the series below:

$$1 + \left(\frac{1}{2}\right) + \left(\frac{1}{4} + \frac{1}{4}\right) + \left(\frac{1}{8} + \frac{1}{8} + \frac{1}{8} + \frac{1}{8}\right) + \left(\frac{1}{16} + \frac{1}{16} + \frac{1}{16} + \frac{1}{16} + \frac{1}{16} + \frac{1}{16} + \frac{1}{16} + \frac{1}{16}\right) + \cdots$$

Which series will have the larger sum? Why?

Now the second series must diverge. Why?

So the first series must diverge. Why?

Homework

1) Do all the italicized in the *Read and Study* section. Pay particular attention to understanding the proofs.

2) Decide whether each of the following statements is True or False. In each case, explain your answer.

   a) If a series does not diverge, then it must converge.
   b) If the individual terms of a series eventually get arbitrarily close to zero, then the series must converge.
   c) If a series converges then its individual terms must eventually get arbitrarily close to zero.
   d) $0.3 + 0.03 + 0.003 + 0.0003 + \ldots$ is exactly equal to $\frac{1}{3}$.
   e) $0.000000 \ldots 1 = 0$
   f) $1 - 0.9999\ldots = 0.000\ldots1$.

3) Consider the series discussed in the Read and Study: $\frac{1}{2} + \frac{1}{4} + \frac{1}{8} + \frac{1}{16} + \ldots$
   a) Find a formula for the nth term in the sequence of partial sums.
   b) Prove the limit of that sequence is 1 by finding a formula for the number of terms required for the partial sums to be within a distance of ε of 1.

4) For each series do the following:
   i) Use a spreadsheet to make a graph of the sequence of partial sums s_n. Show enough terms to convey the general behavior of this sequence.
   j) Based on the graph, does the sequence of partial sums converge to a fixed value L? If so what is the value of the limit L?
      - Using $\varepsilon = 0.1$, determine the value N so that s_n is within the distance ε away from L for all n greater than N.
      - Using $\varepsilon = 0.001$, determine the value N so that s_n is within the distance ε away from L for all n greater than N.
   k) If the sequence of partial sums does not converge, explain how you can tell.

   a. $\frac{3}{10} + \frac{3}{100} + \frac{3}{1000} + \frac{3}{10000} + \ldots$

   b. $\frac{2}{3} + \frac{4}{9} + \frac{8}{27} + \frac{16}{81} + \ldots$

5) Does the infinite series $\frac{1}{10} + \frac{6}{100} + \frac{6}{1000} + \frac{6}{10000} + \cdots$ diverge or converge? If it converges, what is its sum? If it diverges, make an argument that it diverges.

6) Give two new (not already in the text) examples of infinite series that diverge, and argue that they indeed do diverge.

7) Here is an example of a series we call an **alternating series**; Leibniz (1646-1716) gave an early proof that it converges to π:

$$\pi = \frac{4}{1} - \frac{4}{3} + \frac{4}{5} - \frac{4}{7} + \frac{4}{9} - \cdots$$

   a) Determine the next two terms of this series.
   b) Explore the sequence of partial sums using a calculator or computer. How many terms do you need to get partial sums within 0.01 of the exact value?
   c) How do you know it is theoretically possible to get the partial sums within 0.01 of the exact value? Explain.
   d) This series shows that the common notion that "there is no pattern to the number π" is merely an artifact of choosing to write π as a decimal. Decimals are numbers written as sum of fractions with denominators that are increasing powers of ten. If we instead write π as sum of fractions with denominators that are the sequence of odd numbers, the number π has a very simple pattern indeed. Take a moment to appreciate the beauty of this.

7. Finding the sum of the series of reciprocals of square numbers is a famous problem in the history of mathematics. It was called the Basel Problem, first posed by Pietro Mengoli in 1650, and finally solved by Leonhard Euler in 1734 when he was just 28 years old, bringing him great fame. Here's the series:

$$\frac{1}{1} + \frac{1}{4} + \frac{1}{9} + \frac{1}{16} + \frac{1}{25} + \cdots$$

   a) Use a spreadsheet to make a graph of the sequence of partial sums. Show enough terms to convey the general behavior of the sequence.
   b) Leonhard Euler was able to prove that this series converges to $\frac{\pi^2}{6}$. That's amazing! (Both that that's the exact number that series sums to, and also that it can be proven!) Using $\varepsilon = 0.01$, determine the value N so that s_n is within the distance ε away from $\frac{\pi^2}{6}$ for all n greater than N (you can use a calculator's decimal approximation for $\frac{\pi^2}{6}$).

Class Activity 5: Rational Numbers and Geometric Series

Bridges would be safer if only people who knew the proper definition of a real number were allowed to design them.

David Mermin

A **real number** can be defined informally to be any (directed) distance. Geometrically the set of real numbers can be modeled as the set of points on the number line. Five is a real number. So is $\frac{1}{5}$. So is 0.067534 and so is ⁻0.12122122212222...

Every real number can be classified as either rational or irrational.

Rational numbers are those real numbers that can be written as a fraction, or *ratio* of two integers $\frac{a}{b}$ (where *b* is not zero). *Note that the word "rational' has 'ratio' in there.*

Irrational numbers are those real numbers that cannot be expressed as the ratio of two integers.

An example of a mathematical object that can involve an infinite process is the **decimal representation** of a real number. For some numbers, the decimal representation can be a finite process, such as ½ = 0.5. For others, this representation can repeat periodically, such as $\frac{1}{3} = 0.333\ldots$ or $\frac{2}{7} = 0.285714285714285714\ldots$, or can continue non-periodically, such as π = 3.1415926535897932384...., or the number 0.12122122212222... *(Is this one really non-periodic? Explain your thinking.)*

1) Begin by doing the following:

   a) Prove that $\frac{2}{7} = 0.285714285714285714\ldots$

   b) Can $\frac{5}{17}$ be written as a periodically repeating decimal? Explain.

2) True or False? A fraction in lowest terms of the form $\frac{a}{b}$ will either terminate or will begin to repeat periodically *at or before the 'bth' decimal place*. Prove your answer. (Take a nice, close look at the $\frac{2}{7}$ example – what are the possible remainders when doing the long division?) Make sure to eventually discuss this question in your class.

3) Write the following decimals as fractions or explain why it is not possible to do so.

   a) 0.432

   b) 0.123123123...

   c) 0.999...

   d) 0.12345678910111213...

   e) 0.12567567567567567...

Read and Study

> *As the finite encloses an infinite series*
> *And in the unlimited limits appear*
> *So the soul of immensity dwells in minutia*
> *And in narrowest limits no limits inhere.*
>
> *Jakob Bernoulli*

A **geometric series** is a series with a constant ratio between successive terms. *Stop and figure out what this means.*

Any geometric series can be written in the form: $a + ar + ar^2 + ar^3 + \ldots$ where a is not zero, and r is the constant ratio between successive terms. (We note that if $r = 0$, we don't have much of a series.) A geometric series converges when $-1 < r < 1$ and diverges for all other values of r.

Before you go any further, test the above claim about convergence for a few values of r and see if it seems to be true. Mathematicians are always testing things. Give them a claim, and they'll immediately get a far-away look in their eyes as they mentally test examples to see if it seems true. Recall that if they find a case where the claim does not hold, they call that a *counterexample*. Make it a habit to model this 'testing behavior' for your students.

According to the claim above, if we have a series: $a + ar + ar^2 + ar^3 + \ldots$ where $-1 < r < 1$, then it has a finite sum, S. Let's see if we can find it.

First we can write
$$S = a + ar + ar^2 + ar^3 + \ldots.$$

Then
$$rS = ar + ar^2 + ar^3 + \ldots. \qquad \text{Why is this?}$$

Now we will cleverly subtract the bottom equation from the top one:

$$S - rS = a \qquad \text{What happened to everything else?}$$

Solving for S, we have:
$$S(1 - r) = a$$

$$S = \frac{a}{(1-r)}$$

So if a geometric series has a sum S, then $S = \frac{a}{(1-r)}$.

Let's test to see if this formula works the way we think it should.

Suppose $S = 0.3 + 003 + 0.003 + 0.0003 + \ldots$.

$$S = \frac{3}{10} + \frac{3}{100} + \frac{3}{1000} + \frac{3}{10000} + \cdots$$

$$S = \frac{3}{10} + \frac{3}{10}\left(\frac{1}{10}\right) + \frac{3}{10}\left(\frac{1}{10}\right)^2 + \frac{3}{10}\left(\frac{1}{10}\right)^3 + \cdots$$

So, S is a geometric series with $a = \frac{3}{10}$ and $r = \frac{1}{10}$. So, based on what we did above,

$$S = \frac{\frac{3}{10}}{\left(1 - \frac{1}{10}\right)} = \frac{1}{3}.$$

Now, we have proved that $0.3333\ldots = \frac{1}{3}$.

In fact *every* periodically repeating decimal representation involves a geometric series in disguise!

So that worked out just the way we hoped it would. But there is something about our argument that bothers us here, and we want it to bother you too. Where in the argument (where we found $S = \frac{a}{1-r}$ above) did we use the fact that $-1 < r < 1$?

In other words, why doesn't this formula work for other values of r? Mathematicians are extremely picky about this sort of thing. *Make sure to bring this up in class.*

Homework

The greatest inspiration is often born of desperation.
Comer Cotrell

1) Do all the things in the *Read and Study* section. Make sure to try to figure out where we used the assumption that $-1 < r < 1$ when we derived the formula for the sum of a geometric series.

2) We claimed that, "Every periodically repeating decimal representation invloves a geometric series in disguise!" Convince yourself that this is true. Use our process to write this one as a geometric series: $0.762762762\ldots$

3) Write each of these series as a single fraction or explain why it is not possible to do so.

   a) 0.150150150 ...

   b) 0.56323232323232...

   c) 0.144214421442...

   d) 0.01001000100001...

4) Write each of these series as a single fraction or explain why it is not possible to do so.

   a) $\frac{9}{10} + \frac{9}{100} + \frac{9}{1000} + \frac{9}{10000} + \cdots$

   b) $\frac{3}{4} + \frac{3}{16} + \frac{3}{64} + \frac{3}{256} + \cdots$

   c) $\frac{3}{4} + \frac{1}{2} + \frac{1}{3} + \frac{2}{9} + \frac{4}{27} + \cdots$

   d) $5 + 3 + \frac{9}{5} + \frac{27}{25} + \frac{81}{125} + \cdots$

5) Use what you have learned about series to prove the following theorem: Any rational number can be written as a terminating or a periodically repeating decimal, and any periodically repeating or terminating decimal is a rational number.

6) Argue that any irrational number can be written as an infinitely non-periodic decimal, and any infinitely non-periodic decimal is an irrational number.

7) Here's a game: Roll a die. If you get a 6, you lose. Otherwise, the value of your die is your special number. Then you continue rolling until either you get your number again (in which case you win) or until you get a 6 (in which case you lose). What is the probability of winning this game?

8) Exploring when decimals will be repeating is part of the middle grades curriculum. Do problems 2-3, 2-4, 2-5 and 2-6 in the Core Connections Course 2 (Grade 7) from the College Preparatory Mathematics (CPM) curriculum. How are students asked to make sense of when a decimal will repeat or when it will terminate?

9) Do problem 2-22 in the Core Connections Course 2 (Grade 7) from the College Preparatory Mathematics (CPM) curriculum. How does Jernome's method for writing repeating decimals as fractions compare with our method for finding the sum of a geometric series?

Class Activity 6: An Irrational Number

He is unworthy of the name of man who is ignorant of the fact that the diagonal of a square is incommensurable with its side.

Plato

The square root of 2 is the length of the hypotenuse of a right triangle with legs of length 1. Therefore you can place $\sqrt{2}$ in one exact spot on the real number line. *Start with a number line like the one below and discuss how you might construct the position of $\sqrt{2}$ with straightedge and compass.*

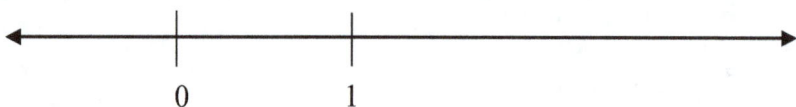

You have probably stated with some confidence that the square root of two is an irrational number. But how do you really know this? How are you convinced? You need to see a proof, of course. Okay. There are many ways to prove that $\sqrt{2}$ is irrational. We'll show you one similar to the proof that the Pythagoreans knew (around 500 BC). *Your group's job is to understand it well enough to explain it to the class.*

Theorem: $\sqrt{2}$ is *not rational*.

Proof: This is another proof by contradiction – so we begin by *assuming* that $\sqrt{2}$ is rational.

Write $\sqrt{2} = \frac{a}{b}$, where *a* and *b* are integers and *b* is not zero. *Why can we do this?*

Now show we can write this equation as $2b^2 = a^2$. Justify your steps.

Now let's write the equation instead in this form:

$$2bb = aa$$

Imagine now that you were to replace *a* in this equation with its prime factorization, and replace *b* with its prime factorization. *Then how many twos would there be on each side of the equation? Stop and think about this before you read any further.*

Well, we don't know exactly unless we knew exactly what the prime factorizations were. But we *do* know that there would have to be an even number of twos on the right side, and an odd number of twos on the left side. *How do we know this?*

But this is impossible. Since prime factorizations are unique (*What theorem are we using here?*), we cannot have a prime factorization with an odd number of twos equaling a prime factorization with an even number of twos. So our initial assumption that $\sqrt{2}$ *is* rational *cannot* be true. Thus $\sqrt{2}$ must be *irrational*.

Read and Study

> *It can be of no practical use to know that Pi is irrational, but if we can know, it surely would be intolerable not to know.*
>
> E. C. Titchmarsh

So just how many irrational numbers are there? In the Class Activity, we proved that $\sqrt{2}$ is irrational. The argument we made can be modified to show that $\sqrt{3}, \sqrt{5}, \sqrt{6}, \sqrt{7}, \sqrt{8}$, and $\sqrt{10}$ are irrational; in fact, the square root of any whole number that is not a perfect square is irrational. We can also prove with a similar argument that $\sqrt[3]{2}$, the cube root of 2 is irrational, and in fact, the cube root of any whole number that is not a perfect cube is irrational. In general, the *n*-th root of any whole number that is not a perfect *n*-th power is irrational. This guarantees the existence of an infinite number of irrational numbers.

Other famous examples of irrational numbers include π and e. While the ancient Greeks (500 BC) could prove that roots like $\sqrt{2}$ are irrational, the proofs that π and e are irrational are attributed to Johann Heinrich Lambert and Leonard Euler, respectively, in the mid-1700's. It is still *unknown* whether $\pi + e$, 2^e or $\pi^{\sqrt{2}}$ are rational or irrational.

In the previous section, we discovered that rational numbers are precisely those whose decimal representation repeats periodically. Thus irrational numbers are precisely those whose decimal expansion is infinitely non-periodic. Using decimal representations, then, we can come up with many more examples of irrational numbers, such as 0.1011011101111011111... and 0.1234567891011121314...

Okay, so we have many examples of irrational numbers, in fact, infinitely many. But just where are these irrational numbers? For example, how many irrational numbers are there between $\sqrt{2}$ and $\sqrt{3}$? How are the irrational numbers distributed on the number line? One way to answer these questions is to think about whether the irrational numbers form a dense set.

An ordered set is said to be **dense** if it has the property that between any two elements there is another element in that set. For example, the set of integers is not dense. That means, between any two integers, there need not be another integer. *Why not? Prove that set of integers is not dense.*

However, the set of rational numbers is dense. That is, given any two rational numbers, there is another rational number between them. One way to show this would be to represent rational numbers as fractions. For example, say you are given the rational numbers ¼ and 1/3.. These two numbers are fairly close together, but can you find a rational number between them? Yes. Lots. One easy way to find one would be to write each fraction with a common denominator. Since $\frac{1}{4} = \frac{6}{24}$ and $\frac{1}{3} = \frac{8}{24}$, one choice for a fraction that is in between would be $\frac{7}{24}$. *Notice that the least common denominator is 12, not 24. Why didn't we use the least common denominator in this case?*

To prove that the rational numbers are dense, we would need to show that this property holds, not just for an example of our own choosing, but for *any* two rational numbers, say $\frac{a}{b}$ and $\frac{c}{d}$, where a, b, c, d are integers, and neither b or d is zero. We will ask you to do this in the Homework.

The neat thing about this is that we now have two rational numbers that are even closer together, say $\frac{1}{4}$ and $\frac{7}{24}$, and we can repeat this process to find a rational number between them as well. *Do this. Find a fraction between $\frac{1}{4}$ and $\frac{7}{24}$.* In fact, we can repeat this process indefinitely! So between any two rationals there is in fact an infinite number of rationals. Thus no matter how closely we zoom in on the number line, we will always be able to find rational numbers there. In other words, they are dense on the number line.

Amazingly, the same is true for irrational numbers. To see that the irrational numbers are dense, we will think of irrational numbers as decimal representations that are infinite and non-periodic. For example, let's find an irrational number between $\sqrt{2}$ and $\sqrt{3}$. Using decimal representations, $\sqrt{2} = 1.4142135...$, and $\sqrt{3} = 1.7320508...$. This allows us to see many numbers that are in between, such as 1.5 or 1.72 (though these are rational numbers). To find an irrational number, we just need make sure that the decimal has an infinite non-periodic "tail". One possibility is 1.5115111511115111115111115...

In this way, we can always find an irrational number in between any two given irrational numbers. So again, no matter how closely we zoom in, we will always be able to find more irrational numbers.

Connections to Teaching

Irrational numbers are part of the Common Core State Standards in Grade 8. Here's an exerpt:

The Number System	8.NS

Know that there are numbers that are not rational, and approximate them by rational numbers.

1. Know that numbers that are not rational are called irrational. Understand informally that every number has a decimal expansion; for rational numbers show that the decimal expansion repeats eventually, and convert a decimal expansion which repeats eventually into a rational number.

2. Use rational approximations of irrational numbers to compare the size of irrational numbers, locate them approximately on a number line diagram, and estimate the value of expressions (e.g., π²). *For example, by truncating the decimal expansion of √2, show that √2 is between 1 and 2, then between 1.4 and 1.5, and explain how to continue on to get better approximations.*

In the homework we will ask you to look at how these standards are met in an example curriculum.

Homework

There is no substitute for hard work.

Thomas Alva Edison

1) Answer all of the questions in italics in the Read and Study.

2) Prove that $\sqrt{3}$ is irrational.

3) Why couldn't we use a similar argument to show that the square root of 4 is irrational? Exactly which step(s) in the proof breaks down in that case?

4) Prove that $\sqrt[3]{2}$ is irrational.

5) Without using the square root key on a calculator or looking it up, find the decimal representation for $\sqrt{2}$ to three decimal places. Explain your thinking. Now do the same with $\sqrt{3}$.

6) Without using the cube root key on a calculator or looking it up, find the decimal representation for $\sqrt[3]{100}$ to three decimal places. Explain your thinking.

7) Which pair of numbers in the set $\{2^3, 3^2, \sqrt{72}\}$ are closest together? Figure this out mentally, without using a calculator.

8) Prove that the rational numbers are dense by finding a rational number in fraction form between $\frac{a}{b}$ and $\frac{c}{d}$.

9) Find an irrational number between $\frac{8}{33}$ and $\frac{49}{198}$.

10) Do problem 2-23 and 2-24 in the Core Connections Course 2 (Grade 7) from the College Preparatory Mathematics (CPM) curriculum. How do these problems prepare students to be ready to distinguish between rational and irrational numbers in Grade 8?

11) Read the introduction to Section 9.2.4 in the Core Connections Course 3 (Grade 8) from the College Preparatory Mathematics (CPM) curriculum, and do problems 9-100 and 9-101. How do these problems meet the Common Core State Standard listed in the Connections to Teaching Section?

12) Do problems 9-107 and 9-108 in the Core Connections Course 3 (Grade 8) from the College Preparatory Mathematics (CPM) curriculum. Make sure you come up with ways to do these problems without using a calculator. What knowledge about numbers and operations do 8th graders need to have to be able to do these problems without a calculator? How do these problems meet the Common Core State Standard listed in the Connections to Teaching Section?

Class Activity 7: Life in Hell

Alice laughed: "There's no use trying," she said; "one can't believe impossible things." "I daresay you haven't had much practice," said the Queen. "When I was younger, I always did it for half an hour a day. Why, sometimes I've believed as many as six impossible things before breakfast."

*Lewis Carroll
Alice in Wonderland*

The following problems are adapted from Smullyan (1992).

1) You are in Hell and the devil has challenged you to a game. He will think of a natural number, and you can have one guess at it each day. (Once he has thought of his number, he will not change it.) If you guess his number, he will let you out. Is there a 'guessing strategy' to follow so that you are guaranteed to eventually get out? (Remember – you have eternity.) Either give a foolproof strategy or explain why there isn't one.

2) What if instead of a natural number he tells you that he is going to pick an integer? (You still get one guess each day.) Is there a strategy this time for eventually getting out? Either give a strategy or explain why there isn't one.

3) What if he chooses a rational number? Either give a strategy or explain why there isn't one.

4) What if he chooses a real number? Either give a strategy or explain why there isn't one.

5) What if he chooses a real number between 0 and 1? Either give a strategy or explain why there isn't one.

Read and Study

If you were to paint the real number line, making all the rational points blue and the irrationals pink. The whole line would look pink.

Robert W. Earles

We know that every real number can be classified as either rational or irrational, and both types of number are dense on the number line. So it sort of seems like the two sets are about the same size (whatever that means). However, some disturbing mathematical results suggest otherwise. For example, it is possible to construct a function that that is continuous (unbroken) at every irrational value of *x* and broken at each rational value of *x*. But the reverse has been proved to be impossible. There can be no function that is continuous at every rational value of *x* and broken at every irrational value of *x*. So the rationals and the irrationals are fundamentally different as sets. But what exactly is the difference?

Enter a 19th century German mathematician named Georg Cantor. He put infinity on a solid logical foundation and described a way to do arithmetic with infinite quantities. His basic definitions are these:

Two sets are the same size (have the same **cardinality**) if each object in one set corresponds with exactly one object in the second set. For example, you have the same size set of fingers on each hand, if you can match each finger on one hand with exactly one on the other. *Try this.*

A collection is **infinite**, if it can be put in **one-to-one correspondence** with a **proper subset** of *itself*. For example, the set of natural numbers is infinite according to this definition because the set of say, even numbers is a subset of the natural numbers and we can make a one-to-one correspondence between the set of natural numbers and the set of even natural numbers.

$$
\begin{aligned}
1 &\leftrightarrow 2 \\
2 &\leftrightarrow 4 \\
3 &\leftrightarrow 6 \\
&\vdots \\
n &\leftrightarrow 2n
\end{aligned}
$$

A one-to-one correspondence is a function and it can often be written as a formula. For example, the function given above between the set of naturals with the set of evens can be written as $f(n) = 2n$.

Prove that the set of integers is infinite using Cantor's definition.

Is the set of real numbers infinite according to this definition? Yes. We can show that there is a one to one correspondence between the open interval (0, 1) and the entire real line. The following diagram may help you think about this. Imagine taking the unit open interval (0, 1), lifting it up and wrapping it around in the shape of a semi-circle. The diagram below shows how you can match-up every point on the line with a unique point on the semicircle. *Make sure you understand this. Put a point P on the semi-circle and then identify the location of point, P' on the real line.* This demonstrates that there is the same number of points on the semicircular arc as there is on an entire unbounded line. There are the same number of points in the interval (0, 1) as there are in the entire real line.

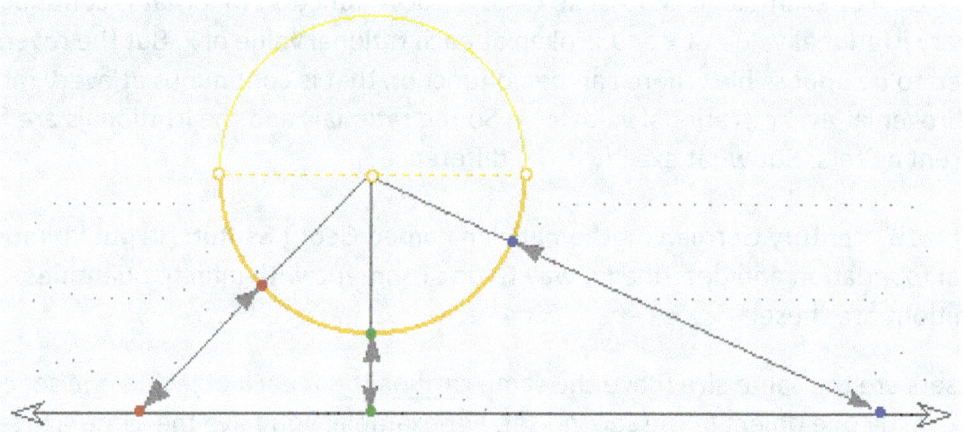

So both the set of natural numbers and the set of real numbers are infinite. But do they have the same cardinality? In other words, is the set of natural numbers the same size as the set of real numbers? Or are they different sizes? If they are different, how many "sizes of infinity" are there?

Infinite sets that are the same cardinality as the natural numbers are called **countably infinite**; these are sets of numbers that can be put into a list (which is just a one-to-one correspondence with the set of natural numbers). We have shown above that the set of even numbers is countably infinite. *Which of the sets in Life in Hell are countably infinite?* It might have surprised you that you could list the rational numbers in Life in Hell. This means that the set of rationals is the same size as the set of natural numbers. Both are countably infinite.

Are there sets that are "bigger" that than the set of natural numbers? Yes. We call these sets **uncountable** (infinite, but not countably infinite). In fact, the set of real numbers is uncountable.

This is an amazing statement! We need to prove it – or at least we need to understand a proof. Below is Cantor's famous proof that the real numbers are uncountable. This is another proof by contradiction. *Fight to understand it*; it's a great argument.

Theorem: The set of real numbers is uncountable.

Proof: (Cantor's Diagonal Argument): We have already seen that the set of real numbers has the same cardinality as the set of real numbers in the open interval (0, 1). We will prove that this unit interval is uncountable and hence the set of real numbers is uncountable.

To do this we will argue by contradiction. We first assume that the open interval (0, 1) *is* countable. That is, we assume it has the same cardinality as the natural numbers, and thus all the real numbers between 0 and 1 can be put in one-to-one correspondence with the naturals.

Next we decide to represent each real number in the interval (0, 1) as a decimal. If there is a one-to-one correspondence with the natural numbers, we can make a list of all of the decimals in the unit interval, which would look something like this:

$$1 \leftrightarrow d_1 = 0.d_{11}d_{12}d_{13}d_{14}\ldots$$
$$2 \leftrightarrow d_2 = 0.d_{21}d_{22}d_{23}d_{24}\ldots$$
$$3 \leftrightarrow d_3 = 0.d_{31}d_{32}d_{33}d_{34}\ldots$$
$$4 \leftrightarrow d_4 = 0.d_{41}d_{42}d_{43}d_{44}\ldots$$
$$\vdots$$
$$n \leftrightarrow d_n = 0.d_{n1}d_{n2}d_{n3}d_{n4}\ldots$$
$$\vdots$$

*Don't let our notation confuse you. First study it. Notice for example that we named the **n**th decimal on the list d_n. Since this is some decimal between 0 and 1, the ones place must be 0. The tenths place is represented by d_{n1} and the hundredths place by d_{n2}, etc. For example. maybe d_n happens to be 0.3925884…. Then d_{n1} = 3, d_{n2} = 9, d_{n3} = 2, etc.*

Note that we are assuming that *all* the real numbers between 0 and 1 are part of the one-to-one correspondence; all of them are on the list and none are missing. Imagine that it is a complete list.

Now, we will show that we can use this list of all the numbers between 0 and 1 to construct a number between 0 and 1 that is not on the list. (*Note that this would be a big logical problem. Bad. Really bad. Mathematicians would rather agree that the set of decimals between 0 and 1 is bigger than the set of naturals than accept that this is possible.*)

Consider the decimal number $x = 0.\ x_1x_2x_3x_4x_5\ ....$, where x_1 is any digit other than d_{11}; x_2 is different from d_{22}; x_3 is not equal to d_{33}; x_4 is not d_{44}; and so on. (And just to be careful, let's not let any of the x_n's be a 0 or a 9 – *Why not? Make sure to ask in class if you can't figure it out*. Mathematicians love to be put on the spot.)

Now, **x** is a decimal number, and **x** is between zero and one, so it must be in our list. But where? **x** can't be first, since its first digit differs from d_1's first digit. **x** can't be second in the list, because **x** and d_2 have different hundredths place digits. In general, **x** is not equal to d_n, since their nth digits are not the same.

So **x** is nowhere to be found in the list. We have proved that if we assume we have a complete list of the real numbers between 0 and 1, we can use the list to construct a number between 0 and 1 that is not on the list. That is a logical contradiction. So the assumption that we can make a complete list of the real numbers between 0 and 1 is false. Thus the set of real numbers between 0 and 1 is uncountable.

Since the set of numbers in the interval (0, 1) is the same size as the set of real numbers, we have also shown that the real numbers are uncountable. The real are numbers are a bigger size of infinity than the natural numbers.

Connections to Teaching

A teacher affects eternity; he can never tell where his influence stops.
Henry Brooks Adams

Philosophically, there are at least two ways of thinking about infinity.

1) Infinity is the *potential* for a never-ending process. Thinking this way, the process of counting the naturals is never complete. For every natural number, there is always one bigger. Thinking this way, 0.999… is never finished, and so it might not make sense to talk about 0.999… as a number at all.

2) Infinity is an *actual* size or an actual completed process. It makes sense to say that there are infinitely many natural numbers. It makes sense to think of having (as an idea) a complete list of all of them. It makes sense to think of 0.999… as being "done." Cantor's work is consistent with this idea, and to understand the real number system, students will need to begin to think like this as well.

As mathematicians, we hold both conceptions and try to evoke the one that is most useful in a given problem situation. Tall (1992) suggests that children's experiences with infinite processes typically strengthen the idea of infinity as *potential* and not the idea of an *actual* infinity.

Homework

No pain. No gain.

Anonymous

1) Do all the italicizing things in the *Read and Study* section.

2) Is the set of irrational numbers countable or uncountable? Explain your reasoning.

3) Prove that the set of perfect square numbers is countably infinite. Find a formula for your one-to-one correspondence.

4) Argue that you can throw away 99% of the natural numbers, and be left with a set that has the same cardinality as the natural numbers.

5) Find a formula for a one-to-one correspondence that shows that the interval [0, 1] has the same cardinality as the interval [-10, 10].

6) Prove the set of real numbers is infinite according to Cantor's definition by finding one-to-one correspondance between the interval (0,1) and the entire real line that uses a common function from high school algebra.

7) Prove that the set of rational numbers has the same cardinality as the natural numbers. (Find a way to list the rationals so they can put in one-to-one correspondence with the naturals.)

8) Explain the big idea of Cantor's Diagonal Argument. What was he proving and what was the structure of his argument?

Class Activity 8:   To Infinity and Beyond!

The essence of mathematics is not to make simple things complicated, but to make complicated things simple.

S. Gudder

Given a set A, the **power set**, denoted **P**(A), is the set of all the subsets of A.

If a set has *n* elements, then how many elements are in its power set? Prove it.

Read and Study

> *The fear of infinity is a form of myopia that destroys the possibility of seeing the actual infinite, even though it in its highest form has created and sustains us, and in its secondary transfinite forms occurs all around us and even inhabits our minds.*
>
> Georg Cantor

We proved in the previous section that the size of the infinity of real numbers is not same as the size of the infinity of natural numbers. It is bigger. So there are at least two different "sizes of infinity", perhaps more. So we would like some way of describing the cardinality of infinite sets that would show that the set of natural numbers and the set of real numbers have different cardinalities.

With finite sets, we are able to assign a single number to the set that represents its cardinality. Is there a symbol we can assign to an infinite set to represent how many elements are in the set? We need to invent a new kind of number, what Cantor called a *transfinite number*.

We define the transfinite number $\aleph_0$ to be the cardinality of the natural numbers. (The symbol $\aleph_0$ is pronounced *aleph naught*. Aleph is the first letter of the Hebrew alphabet.) So if someone asked you how many natural numbers there are, you could just say $\aleph_0$. *How many prime numbers are there? How many integers are there? How many rational numbers are there?*

Now what about the cardinality of the real numbers? Cantor's diagonal argument proves that the cardinality of the real numbers cannot be $\aleph_0$. It must be greater than $\aleph_0$. Just as he invented $\aleph_0$, he created another transfinite number to measure the size of the real numbers. He used c to denote the cardinality of the real numbers. (The letter c stands for *continuum*.) Then we know that

$$\aleph_0 < c$$

Now that we have two different transfinite numbers, it is natural to ask if there are more? Yes. Lots. This is the result of another important theorem proved by Cantor, which we will show you now.

Theorem (Cantor): For any set A, the cardinality of its power set **P**(A) is greater than the cardinality of A.

> **Proof:** We have already seen in the Class Activity that this is true for finite sets. Now we must prove that this is true for infinite sets as well. So we will assume A is an infinite set.
>
> It is easy to see that the cardinality of **P**(A) cannot be less than the cardinality of A. *Why not?* So all that remains is to prove is that the cardinalities cannot be equal.

We will prove this by contradiction. That is, first we assume that the cardinality of A is the same as the cardinality of **P**(A). Then there exists a one-to-one correspondence between these two sets. That means for each element in A, there is a corresponding element of the power set (that is, a subset of A). Now we will define a set that will lead to our contradiction. This set is very interesting, but it takes some work to understand it. Ready?

Define the set B to be the set of each and every element of the original set A that is *not* a member of the subset with which it is matched. *Read through the definition again. Fight to understand it.*

To better understand this set B, let's consider this example:

Suppose A = {a,b,c,d,e,f,g,....}. Then a possible one-to-one correspondence with **P**(A) could be:

Elements of A		Elements of $\mathbf{P}(A)$
a	$\leftrightarrow$	$\{b, c\}$
b	$\leftrightarrow$	$\{a, b, c, d\}$
c	$\leftrightarrow$	A
d	$\leftrightarrow$	$\{e, f, g, \ldots\}$
e	$\leftrightarrow$	$\{a, c, e, g, \ldots\}$
f	$\leftrightarrow$	$\{\}$
g	$\leftrightarrow$	$\{d\}$
$\vdots$		$\vdots$

Now let's use the definition of the set B to decide which elements of A in the set B. First let's consider the element *a*. Is *a* a member of the subset with which it is matched? No. Thus *a* is in set B.

Is *b* a member of the subset with which it is matched? Yes. Thus *b* is not the set B.

Continuing this process for each of the elements of set A given in this example you will see that of the elements listed, *a, d, f* and *g* are in the set B, while *b, c, e* are not. Thus B = {a,d,f,g, ...}. Don't take our word for it. Convince yourself that this is right.

Now that we understand this set B, we see that it is a well-defined set. Furthermore, the set B must be in the power set of A. *Why must B be in* **P**(A)?

Since *B* is in **P**(*A*), it must appear somewhere in the one-to-one correspondence, and it must have some element in the set *A* that corresponds to it. Let's call that element *x*.

$$
\begin{array}{ccc}
a & \leftrightarrow & \{b, c\} \\
b & \leftrightarrow & \{a, b, c, d\} \\
c & \leftrightarrow & A \\
d & \leftrightarrow & \{e, f, g, \ldots\} \\
e & \leftrightarrow & \{a, c, e, g, \ldots\} \\
f & \leftrightarrow & \{\} \\
g & \leftrightarrow & \{d\} \\
\vdots & & \vdots \\
x & \leftrightarrow & B \\
\vdots & & \vdots
\end{array}
$$

Now we ask: is the element *x* in the set *B*?

Well, there are only two possibilities: either *x* is in the set *B*, or it isn't. But if you think about each of these two cases, you will see that neither of these is possible.

Why is it impossible that x is in B?

Why is it impossible that x is not in B?

So we have found a *logical* contradiction. If the set *A* and its power set **P**(*A*) had the same cardinality, then we could find a one-to-one correspondence between them. But then we would be able to define this set *B*, which could have no element of *A* to correspond to it. So such a one to one correspondence is not possible. So *A* and **P**(*A*) cannot have the same cardinality. **P**(*A*) must be bigger.

Going back to our transfinite numbers, we have decided to let $\aleph_0$ be the cardinality of the set of natural numbers $\mathbb{N}$, and *c* represent the cardinality of the set of real numbers $\mathbb{R}$. We know that these are two different transfinite numbers, and that $\aleph_0 < c$. Two questions now arise: 1) is there a transfinite number *between* $\aleph_0$ and *c*, and 2) is there a transfinite number *greater* than *c*? The second question is easily answered. We will let you think about it now and then ask you again in the homework section.

Let's look now at the first question: is there a transfinite number between $\aleph_0$ and *c*? Well, if there were, one candidate would be the cardinality of the power set of the natural numbers or **P**($\mathbb{N}$), since that must be larger than the cardinality of the natural numbers. But, in fact, it can be shown that the cardinality of **P(ℕ)** is precisely the cardinality of the real numbers. That is, there is a one-to-one correspondence between the set of real numbers and the set of subsets of natural numbers. One way to see this is to think about the decimal representation of real numbers, as you will be asked to do on the homework.

The question still remains whether there is a transfinite number between $\aleph_0$ and c. Cantor's Continuum Hypothesis is that the answer is "no". According to this hypothesis, the next largest set after the set of natural numbers is the set of real numbers; there are no sizes in between. Cantor was unable to prove his hypothesis. Amazingly, it has been since proved that the Continuum Hypothesis is not provable to be true or false, given the axioms of arithmetic. That is, the Continuum Hypothesis being true is consistent with our axioms, but so is the Continuum Hypothesis being false.

Let's end this section by looking at some uncountable sets that we are familiar with from geometry, and think about their cardinality. First consider the simple line segment. Viewing a line segment as a set of points, by putting this line segment on the *x*-axis we see that a line segment is really just an interval of real numbers. We know that this set is uncountably infinite and has cardinality c, the transfinite number which is the cardinality of the real numbers. *How do we know that an interval has the same cardinality as the whole real number line?*

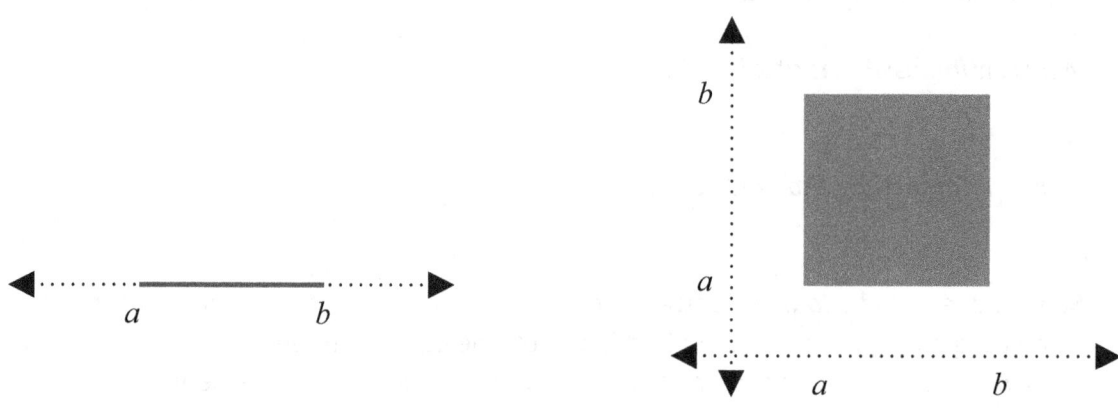

Now let's consider a square, or more precisely a filled-in square. This is a two dimensional object. If we consider the set of points in a square, it *seems like* this set should have a larger cardinality than the set of points in a line segment, a one-dimensional object. So intuition suggests that the cardinality of the square should be *larger* than c. (Perhaps one might guess that the square could might have the cardinality of the power set of real numbers, which we know is larger than c.)

However, as you will see in the homework section below, Cantor proved that a square has the same cardinality as a line segment, so in fact the square also has cardinality c. In other words, there are as many points in a square as there are along just one of its sides! Cantor showed with similar arguments that a cube has the same cardinality as a line segment, and in general, the cardinality of *n*-dimensional space is the same as the cardinality of the real line.

Another way to see that a one-dimensional set and a two-dimensional set can have the same cardinality is by way of a space-filling curve, which is a one-to-one correspondence between a line segment and a region of space. One way to generate a space-filling curve is through an infinite iterative process like the ones used to create fractals. Shown below is a famous example of a space-filling curve, called the Hilbert Curve, or at least the first several iterations. At each step, we have a single bent-up line segment that starts at the bottom left and ends at the bottom right. In the infinite limit, we have a curve that fills up the entire square!

Image in public domain. Retrieved from http://commons.wikimedia.org/wiki/File:Hilbert_curve.png

Homework

No one shall expel us from the paradise which Cantor created for us.
David Hilbert

1) Do all the things in the *Read and Study* section. In particular, make sure you can explain why it is impossible for *x* to be in set *B*, and explain why it is impossible that *x* not be in set *B*.

2) A pizzeria offers 10 different toppings for its pizza. In the "build-your-own" pizza option, you get to choose as many toppings as you want, from no toppings up to all ten. How many different build-your-own pizzas are possible?

3) a) Consider a set *A* with two elements, say $A = \{a,b\}$. List all of the elements of **P(P(A))**.
   b) If the set *A* has 5 elements, how many elements does **P(P(A))** have?

4) Is there a cardinality greater than the cardinality of the set of real numbers? Explain.

5) True or False: If A is a proper subset of B, then the cardinality of A must be less than the cardinality of B. Explain your thinking.

6) To show that the cardinality of the unit interval is the same as the cardinality of the unit square, Cantor argued as follows. Of course he needed to find a one-to-one correspondence between points in the unit square and points in the unit interval. We will describe one way to set up such a correspondence.

Let (x, y) be any point in the unit square. Then x is some number in the unit interval [0, 1], and y is also some number in the unit interval [0, 1]. Suppose x has a decimal representation $x = 0.a_1 a_2 a_3 a_4 \ldots$ and y has a decimal representation $y = 0.b_1 b_2 b_3 b_4 \ldots$. Then we can choose to correspond (x, y) with the number $z = 0.a_1 b_1 a_2 b_2 a_3 b_3 a_4 b_4 \ldots$, which is in the unit interval [0,1]. Using the one-to-one correspondence described above, to which point in the unit square do the following points in the unit interval correspond to?

a) 1/8 = 0.125
b) π/4 = 0.785398163397....
c) 2/3 = 0.666...

7) See if you can find a one-to-one correspondence between the real numbers in the unit interval [0,1] and the power set of the natural numbers. (Hint: use decimal representations).

Chapter Two: Functions and Modelling

Class Activity 9:   Making Models

Brady's First Law of Problem Solving:
When confronted by a difficult problem, you can solve it more easily by reducing it to the question: How would the Lone Ranger have handled this?
Brady

Sketch a graph that could represent the given situation. Make sure that you label and provide a consistent scale on each axis. Be prepared to defend why you think your graph is reasonable.

1) Sketch average daily high temperature (in your city) over the course of one year.

2) A poorly paced distance runner running a 10k race starts off quickly, tires and slows to a walk, then sprints to the finish.
    a) Sketch a graph of the runner's total *distance* covered vs. time.
    b) Sketch a graph of the runner's *speed* vs. time.

3) Suppose an epidemic strikes the United States and everyone is vulnerable. Sketch a graph of the total number of people who have been infected as a function of time.

4) A swimmer swims a 200-meter race in a 50-meter pool. She dives from the starting block, and then swims laps in her lane. She swims at a constant rate, except that she has to rest for a few seconds at 150 meters. Sketch a graph of the distance from the starting block as a function of time.

5) You plan to have fresh fruit pies at your graduation party. Sketch a graph showing the relationship between the number of people you invite to your graduation party and how many pies you should bake.

Read and Study:

> ... the function concept is anything but an extension or elaboration of previous number concepts – it is rather a complete emancipation from such notions.
>
> W. A. Schaaf
> Mathematics and World History

A **function** is a rule that assigns elements from a set A to elements of another set B in such a way that each element in A (called the **domain**) is assigned to exactly one element of B (called the **range**).

You are probably most used to functions where the domain and range are sets of numbers, and the rule is some algebraic formula, such as $y = 3x + 7$, where x is a number in the domain and y is it's corresponding number in the range. But functions can be expressed in many ways. Here are two examples of functions where the domain and range are not sets of numbers and the rule is not an algebraic formula.

1) Set A: All children.
   Set B: All biological mothers.
   Rule: Each child is assigned to his or her biological mother.

   (Is the rule that assigns each mother to her child a function too?)

2) Set A: {a, b, c}
   Set B: {yellow, green}
   Rule: $a \rightarrow$ yellow, $b \rightarrow$ yellow, $c \rightarrow$ green   (Note: It's okay that two elements in A get assigned to the same element in B.)

Go back to the previous class activity and decide whether each scenario behaves like a function. In each case, which set is the domain and which set is the range?

A very powerful way to think of a function is the **machine concept,** in which you can think of a function as a machine that accepts inputs and spits out outputs. Inside the machine is the rule that considers each input and decides which output to produce from that input.

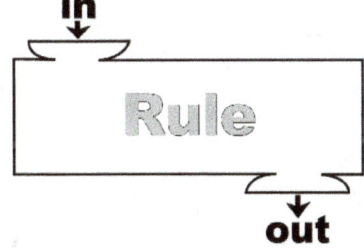

With this viewpoint, we can think of the domain as the set of all of the inputs in to the function, and the range as the set of all of the outputs out of the function.

A function is also a kind of relationship between two variables. The variable whose values are elements of the domain is called the **independent variable** for the function, and the variable whose values are elements of the range is called the **dependent variable** for the function. *Why does it makes sense to call the domain variable "independent" and the range variable "dependent"?* In standard notation, x is used as an independent variable, and y as the dependent variable. Using these variables, a simple machine diagram for a function is as shown.

This functional relationship between x and y is represented in the **function notation** $y = f(x)$. Let's talk a little bit about notation. Mathematicians pay attention to and value notation. A good notation can embody mathematical relationships and aid in computations, and allow us to keep our thinking clear and be able to make precise arguments. This function notation is a good example. When we look at the linear equation $y = 3x - 7$, we can recognize that the variable y is a function of the variable x. To show this feature, we can write the equation using function notation, as $f(x) = 3x - 7$. Here f represents the function.

The familiar equation $y = f(x)$ means that the values of the variable y are the outputs of the function f with inputs x. When we write $f(x)$, we can be referring to either to the function f itself or the values y taken by the function. It's important for you to be able to help your students understand this subtle distinction.

In cases where the domain and range are sets of real numbers then we can make a **graph** of the function. The standard convention when making the graph of a function is to put the independent (domain) variable on the horizontal axis, and the dependent (range) variable on the vertical axis. Think of a graph as a sprinkling of points (x,y). In this case, each point has the height from the range (the "y-value") that is related to the element of the domain (the "x-value"). Take a look at a graph of the function defined by the following equation:

$$y = -11x^2 + 66x, 0 \leq x \leq 6$$

62

Notice a couple of things: First, see how we have specified a domain for the function. We said the relationship is valid for *x*-values from 0 to 6. If a domain is not explicitly stated, then mathematicians usually assume the biggest domain that makes sense for the situation. The range of the function is the set of all possible outputs using elements of the domain as inputs.

Now, *study how the graph illustrates the relationship*. (Remember that we are thinking of this graph as a sprinkling of points for now.) For example, the point above *x* = 2, has a height (or *y* value) of 88. Note that we can tell it's a functional relationship because above (or below in some functions) each *x*-value there is only one *y*-value. *What is the maximum value obtained by this function on its domain (if any)? Explain. What about the minimum value on its domain? What is the range of this function?*

Functions between numerical variables can be represented in several ways: in words, by numeric tables, with an algebraic formula, and with a graph. Let's see all these representations for a specific example: First, the **words**: Suppose that I'm earning 12% annual interest in my checking account and I put $100 in there for eight years. Let's consider the amount of money in my account as a function of time.

Here is that data as a **numeric table**:

Year	Amount
0	$100.00
1	$112.00
2	$125.44
3	$140.49
4	$157.35
5	$176.23
6	$197.38
7	$212.07
8	$247.60

Do the calculations to check that this data is correct. Now, look at the ratios of amounts of money in my account for constant changes in time. For example, compute 112/100 and 125.44/112 and 140.49/125.44 etc. What do you see? This is the hallmark of exponential growth, which we will study in more detail in the next section.

Here is the amount of money in my account captured by an **algebraic model**:

$$y = 100 \, (1.12)^x \quad (\text{or} \quad y = 100 \cdot 2^{(.1635x)} \quad \text{or} \quad y = 100 e^{(.1133x)})$$

Check that these are all the same function and check that they give the above table values. Do you see where that 1.12 comes from?

Finally, here is a **graph** showing the relationship between time (on the *x*-axis) and the money in my account on the *y*-axis. (Think of a graph as a picture of the numeric or algebraic models.)

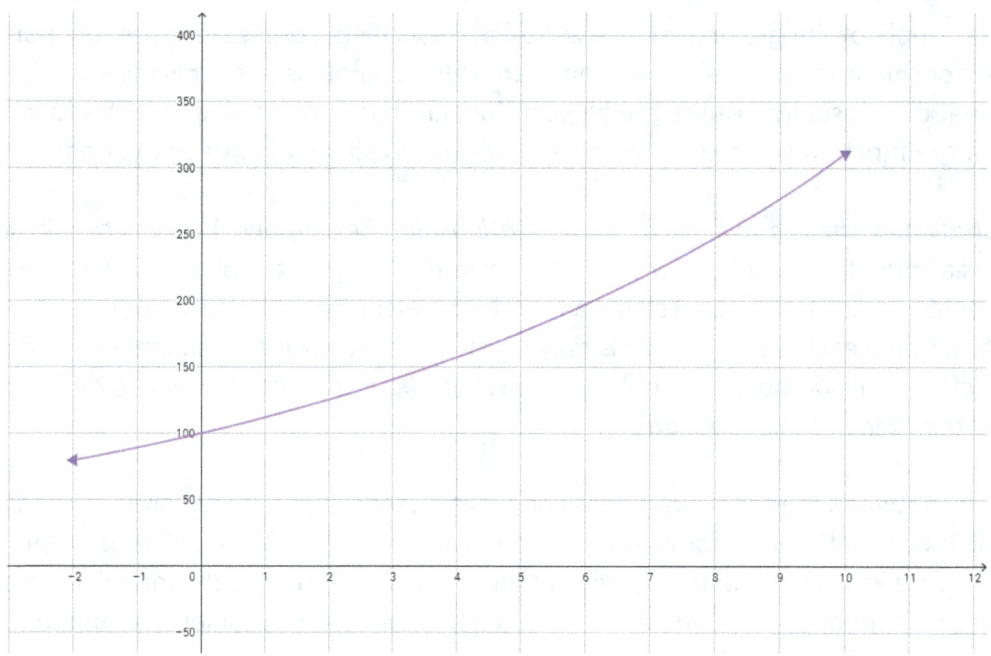

In what ways does the algebraic model fit the real-life description of the situation? In what ways does it fail to fit the real-life situation?

We note that the domain of the algebraic model is bigger than that of the real-life situation. We are interested only in the piece of the graph starting at $x = 0$ and only in amounts of money rounded to the nearest penny. We say that the real-life situation has **discrete** outputs (money to the nearest cent) whereas the theoretical model has **continuous** outputs (all positive real numbers form a continuum).

But what's the big deal with functions? Why do mathematicians like them so much? Here's the secret: functions are powerful models for many real-life phenomena. So if we can look at something that naturally has a relationship in which there is only one *y* for each *x* (like the situations in *Making Models* above), then we might be able to capture that functional relationship *algebraically* (by this we mean *with a formula*) and then we can study the phenomenon by studying the **algebraic** model for the relationship.

Read that paragraph again. It's a big deal.

Connections to Teaching:

> *I am not a teacher, but an awakener.*
> *Robert Frost*

Function. It is a fundamental mathematical idea. BIG. REALLY BIG. And the idea of function appears all through the middle grades curricula. O'Callaghan (1998) outlines a framework for thinking about student learning about functions. The framework has four competencies:

1) **Modeling**: This means translating a problem situation to a mathematical representation of the situation. Typically the representation is an equation, a table, or a graph.

2) **Interpreting**: This means taking a mathematical representation and making sense of it in terms of real-life situations. In a sense, it is the reverse of modeling.

3) **Translating**: This is the ability to move between the various mathematical representations. For example, understanding how changing a constant in an equation affects the look of a graph or knowing how different types of growth shown in a table can be modeled graphically.

4) **Reifying**: This means creating a mental *object* from what was initially viewed as a *process*. For example, when children first start graphing, it is a *process*. They look up each input, find the output and put the point on the (*x*, *y*) plane. However, eventually children will begin to think of the graph itself as a thing (an object) that can be "acted upon." The graph could be shifted, reflected, inverted... That is reifying. Reification is a hard idea. Make sure to discuss it in class.

As you work on problems in this section, think about which competencies are most relevant for each problem, and think about how you might assess your students' understandings of function using this framework.

An introduction to the ideas of *function* and *representations of functions* is a primary focus of the middle grades curricula. These ideas typically are introduced through real-life situations involving types of growth or tracking changes in a variable such as graphing the temperature over time or the distance traveled over time. The focus is often on having students picture the situations graphically.

Homework:

A problem clearly stated is a problem half solved.

Dorothea Brande

1) If you haven't already done so, go back and answer all the questions in italics in the Read and Study section.

2) Give some reasons why the relationship between a domain of the set of your classmates and a range of the set of their favorite colors may not be a function.

3) Algebaic functions typicllay perform a sequence of several operations on the input to determine the output. For each sequence of operations below, do the following:
   - Find $f(4)$
   - Find $f(-2)$
   - Write an algebraic formula for $f(x)$
   - Make a graph of the function $f(x)$
   - Specify the domain and range of the function $f(x)$.

   a. Let $f(x)$ be the function that: adds 1, then takes the reciprocal, then multiplies by 2, then squares.

   b. Let $f(x)$ be the function that: multiplies by 2, then squares, then adds 1, then takes the reciprocal.

   c. Let $f(x)$ be the function that: takes the reciprocal, then adds 1, then squares, then multiplies by 2.

   d. Let $f(x)$ be the function that: squares, then multiplies by 2, then takes the reciprocal, then adds 1.

4) The table shows the number of rabbits living in a field over the course of several months. The number of rabbits listed is the population observed on the first day of each month.

Month	Number of Rabbits
April	230
May	278
June	337
July	407
August	500

   a) At what *rate* is the population of rabbits growing? Explain.

b) Does it make sense that na animal populations might grow like this? Explain.
c) If next year the field starts with only 40 rabbits on April 1, how many would you expect there to be by August 1?
d) Find an algebraic model that fits this data.
e) Is this a discrete or continuous function? Explain.

5) One way that Core Connections Course 3 (Grade 8) of the College Preparatory Mathematics (CPM) curriculum introduces the concept of functions by asking students to find a rule that would generate the data in an "In-Out" table. Read the Introduction to Chapter 3, then do problem 3-3.
   - Find rules that fit the data in each of the tables a)-f) in this problem.
   - Which of the rules you found were linear functions and which are non-linear?
   - Are there only one correct answer to these problems? Explain.
   - Notice that the x-values (inputs) are not listed in any particular order in the table. Why do you think the authors did this? How might this help students to focus on the concept of function? Be sure to discuss this in class.

6) In Core Connections Course 3 (Grade 8) of the College Preparatory Mathematics (CPM) curriculum, the concept of function is defined in problems 8-117, 8-118, and 8-119.
   - Answer all the questions posed in those tasks.
   - In which aspects of these tasks are students asked to reason deductively? Explain.
   - Thinking about O'Callaghan's competencies, in which part(s) of these tasks are students translating?

7) Sketch a reasonable graph of the horizontal distance that a cannonball travels (i.e., the range of the cannonball) as a function of the angle of elevation at which it is launched. Show angles from zero degrees (horizontal to the ground) to 90 degrees (perpendicular to the ground).

8) Suppose water is poured into each container below at a constant rate. For each, sketch a graph of the height of the water in the container as a function of time.

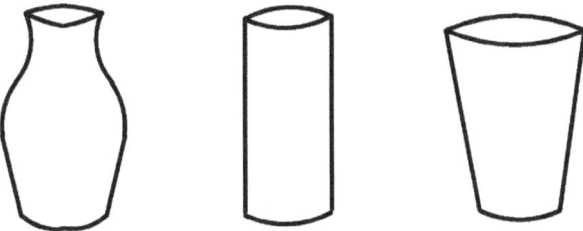

9) The graphs below shows distances (*y*) vs time (*x*)
   - Write a Science Fiction story to fit each of the graphs shown below.
   - Why do you think we specified a *Science Fiction* story?
   - Which of the below graphs are graphs of functions and which are not? Explain.

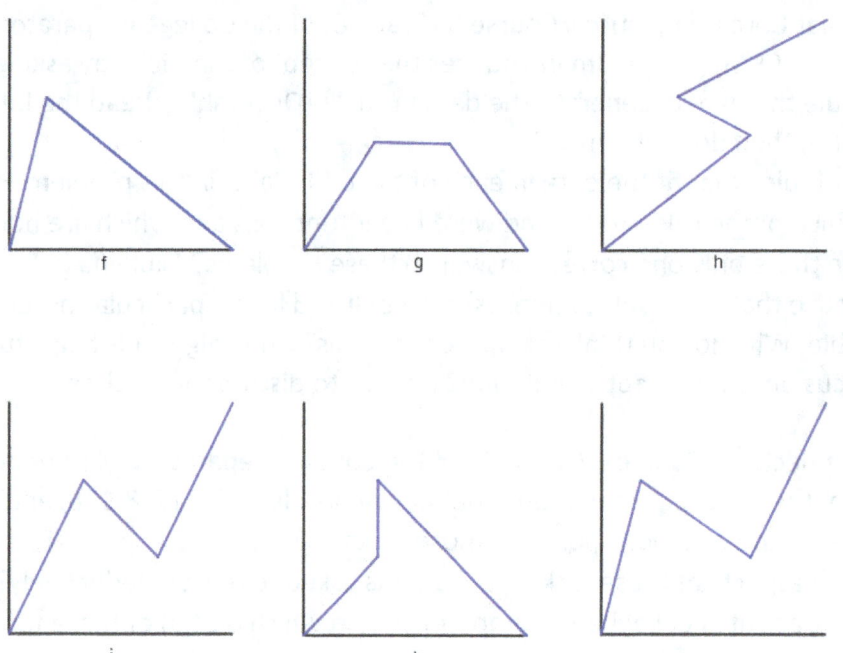

Class Activity 10:   Tides, Rabbits, and Falling Objects

The sky is no longer the limit.

Richard M. Nixon

Below are three sets of mystery data. One of these sets represents the height of water (in feet) in a tidal basin over the course of a day. Another set of data represents the number of rabbits in a field over the course of a year. The third data set represents the height of a ball (in feet) that is dropped from a helicopter as a function of time.

Table I		Table II		Table III	
t	y	t	y	t	y
1	20	1	1984	1	48
3	32	3	1856	3	56
5	51	5	1600	5	48
7	82	7	1216	7	32
9	131	9	704	9	24
11	210	11	64	11	32

1) Decide which table represents each scenario. Then, decide on reasonable units on the time variable in each case.

2) Which of these data sets could be modeled best by an exponential function? How can you recognize exponential growth or decay *from a table*?

3) Which of these data sets could be modeled best by a quadratic function? How can you recognize quadratic growth *from a table*?

4) Which of these data sets could best be modeled by a trigonometric function? How can you recognize trigonometric behavior *from a table*?

5) Find an algebraic model for each of these functions.

6) Use your models to answer the following questions.
   a) How many rabbits were there at time zero?

   b) Assuming growth continues in the same way, how many rabbits do you expect at time $t = 20$?

   c) Is the tide data from our planet? Explain.

   d) How high was the ball when it was dropped? When was it falling fastest?

Read and Study:

Your pain is the breaking of the shell that encloses your understanding.
Kahil Gibran

As we have discussed earlier, **modeling a problem** means finding an algebraic (or sometimes a numeric or graphic) representation of a function that captures the essential elements of the situation. Mathematicians have made up names and notations to represent many types of functional relationships; we'll call these **standard function forms**. If a problem can be modeled by a standard type, we can go ahead and use the appropriate form and fill in the constants to fit that exact situation.

Functions are all about change. If we change the input, we change the output. (Except for constant functions, in which the output is constant no matter what the input is. But this is kind of a boring type of function. So let's just say that all interesting functions are all about change.) Often, if we are lucky, we can see a **pattern** in the way that a function is changing. For example, in Table I from the Class Activity, we can see a pattern in how y changes with respect to t. Namely, as t increases by two, y increases by *multiplying* by 1.6 (or we could say y increases by a factor of 1.6). If instead we looked at the change for every *one* unit in t, we would find that as t increases by one, y increases by multiplying by 1.2649...This pattern holds throughout the entire table. We have seen this phenomenon in the checking account example in the previous section. Each year the money in the account increased by a factor of 1.12.

This kind of change, where for a fixed increase in the independent variable results in a *multiplying* the dependent variable by constant factor, is called **exponential change**. Here there are constant *ratios* between successive y values for constant changes in x. Functions that have this pattern can be modeled using an **exponential form.** For example, $y = kb^x$ is a standard exponential form. Here b is called the base. *Can the base be zero? Can the base be negative? Explain what happens in each case.* To model a function that changes exponentially, we can figure out the appropriate k and b and then use the model $y = kb^x$ to answer questions about the specific situation.

In contrast, with **linear change**, a fixed increase in the independent variable results in *adding* a constant to the dependent variable. Linear change is characterized by constant **differences** in population for constant changes in time. The appropriate model is a **linear form** such as $y = mx + b$.

It is very informative to look at how similar exponential and linear functions are. In both cases, the function can be thought of as changing at a constant rate. The key is that in exponential functions this is a constant *multiplicative* rate, while for linear functions this is a constant *additive* rate. This similarity shows up in the standard function forms as well. In the standard linear form $y = mx + b$, the slope m represents the constant additive rate of change and the y-intercept b is the "initial" value of the function when x is zero. In other words, this function starts out with a value of *b*, then *adds* on *m* a total of *x* times. In the standard exponential form $y = kb^x$, the base *b* represents the constant multiplicative rate of change, and the *y* intercept *k* is the "initial" value of the function when *x* is zero. In other words, this function starts out with a value of *k*, then *multiplies* by *b* a total of *x* times.

Sometimes you can see a pattern where the dependent variable does not change linearly (by adding a constant amount, but by an amount that itself grows linearly. This is the pattern that characterizes **quadratic growth**. Here there are constant **second differences** in the dependent variable. Make sure to talk about this in class. Quadratic growth can be modeled by a function in **quadratic form**, such as $y = ax^2 + bx + c$, or $y = a(x - k)^2 + h$. In the homework you will be asked to explore how changes in the parameters *a*, *k*, and *h* change the graph of a quadratic function.

If a situation exhibits periodic behavior (like a tide or a pendulum), it might fit a **trigonometric form**, such as: $y = a \sin[b(x - c)] + d$. Note that the cosine or tangent may be more appropriate for some periodic behaviors. In the homework you will be asked to explore how changes in the parameters *a*, *b*, *c* and *d* change the graph of a trigonometric function.

Once you have identified an appropriate function form to model your situation, its time to fit your model to the data by specifying the parameters in the form. Be sure to discuss the various techniques for doing this in class. You can also ask your instructor to show you how to ask a computer program or graphing calculator to fit an algebraic model to your data.

Connections to Teaching

Recognize reasoning and proof as fundamental aspects of mathematics.
NCTM, Principles and Standards

When finding a rule to fit a numerical pattern, are two important ways of thinking to distininguish. The first is **recursive thinking,** in which one relates a value (such as a number of rabbits) to its previous value (the number of rabbits in the previous month) In contrast, **functional thinking** is relating a value (such as the number of rabbits) to the value of another variable (such as the number of months). In our experience, students tend see recursive relationships easier, and struggle in finding functional relationships. Often recursive thinking can be used as a starting point to finding functional formulas, like you likely did when finding a function formula for the population of rabbits, but sometimes it requires a shift to a diferente approach.

Consider an example from a classroom in which seventh-grade students were studying "figurate" numbers (drawn from classroom observation and partially described in Malloy [1997]). Reproduced from *Principles and Standards for School Mathematics*. National Council of Teachers of Mathematics (2000), pg 262-233). *Read the first part. Stop when you get to the triangular arrays and figure out the 100th term in the sequence for yourself. Then argue that you are correct using deductive reasoning.*

The teacher began by explaining triangular numbers and then asked the students to generate representations for the first five triangular numbers. The students visualized the structure of the numbers to draw successive dot triangles, each time adding at the bottom a row containing one more dot than the bottom row in the previous triangle (see fig. 6.31). Next the teacher asked the students to predict (without drawing) how many dots would be needed for the next triangular number. Reflecting on what they had done to generate the sequence thus far, they quickly concluded that the sixth triangular number would have six more dots than the fifth triangular number. These students were engaged in recursive reasoning about the structure of this sequence of numbers, using the just-formed number to generate the next number. This approach was repeated for several more "next" numbers in the sequence, and it worked well.

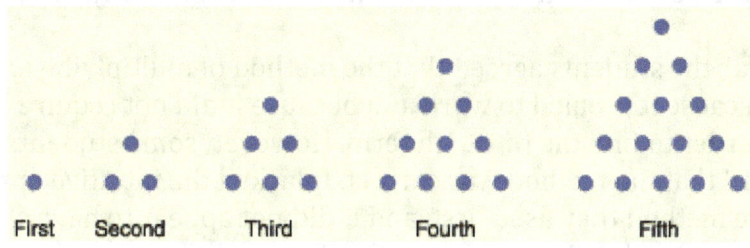

Fig. **6.31.** First five triangular numbers

The teacher then asked the students to find the 100th term in the sequence. Most students knew that the value of the 100th term is 100 more than the value of the 99th term, but because they did not already know the value of the 99th term, they were not able to find the answer quickly. The teacher suggested that they make a chart to record their observations about triangular numbers and to look for a pattern or a relationship to help them find the 100th triangular number. The students began with a display that reflected what they had already observed (see fig. 6.32). They examined the display for additional patterns. Tamika commented that she thought there was a pattern relating the differences and the numbers. She explained that if the consecutive differences are multiplied, the product is twice the number that is "between" them in the display; for example, the product of 4 and 5 is twice as large as 10.

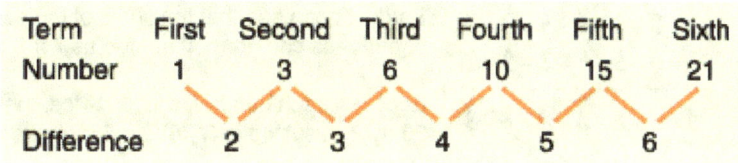

Fig. **6.32.** Triangular numbers

Stop reading now, figure out the 100th triangular number, and make your argument.

Now, read the rest to see how the middle grades class argued the result. This is the sort of conversation that you must strive for with your future students. They are capable of making these arguments.

> The teacher asked the students to check to see if Tamika's observation was true for other numbers in the display. After they verified the observation, the teacher asked them to use this method to find the next triangular number. Some students were unable to see how it could be done, but Curtis used Tamika's observation as follows: "Using Tamika's method, the seventh number is (7)(8)/2, which is 28." Several students checked this answer by using the recursive method of adding 7 to the sixth triangular number to find the seventh triangular number (21 + 7 = 28). The teacher then asked the students to check Tamika's method for the next few triangular numbers to verify that it worked in those instances. She next asked if Tamika's method could be used to find the 100th triangular number. Darnell said, "If Tamika is right, the hundredth triangular number should be (100)(101)/2."
>
> In general, the students agreed that the method of multiplying and dividing by 2 was useful because it seemed to work and because it did not require knowing the nth term in order to find the (n + 1)th term. However, some students were not convinced that the method was correct. It lacked the intuitive appeal of the recursive method they used first, and it did not appear to have a mathematical basis. The teacher decided that it was worth additional class time to develop a mathematical argument to support Tamika's method. She began by asking students to notice that each triangular number is the sum of consecutive whole numbers, which they readily saw from the dot triangles. Then the teacher demonstrated Gauss's method for finding the sum of consecutive whole numbers, applying it to the first seven whole numbers. She asked the students to add the numbers from 1 to 7 to those in the reversed sequence, 7 to 1, as shown in figure 6.33, to see that the seventh triangular number—1 + 2 + 3 + 4 + 5 + 6 + 7—could also be expressed as (7)(8)/2. After the students completed this exercise, the teacher asked them to express the general relationship in words. They struggled, but they came up with this general rule: If you want to find a particular triangular number, you multiply your number by the next number and divide by 2. The students wrote the rule this way: (number)(number + 1)/2.

1 + 2 + 3 + 4 + 5 + 6 + 7 7 + 6 + 5 + 4 + 3 + 2 + 1	Students can see that the sums of the pairs of addends can be represented as 7 × 8, or 56.
8 + 8 + 8 + 8 + 8 + 8 + 8	Because each number is listed twice, they divide 56 by 2, resulting in (7)(8)/2 = 56/2 = 28.

Fig. **6.33.** Gauss's method—sum of first 7 triangular numbers

Homework:

I believe that good things come to those who work.
Wilt Chamberlain

1) Do all the things in italics in the Read and Study section.

2) Explain what aspect(s) of the students work in the triangular number task in the Connections section is using recursive thinking and what aspect(s) is using functional thinking?

3) There are 5000 bacteria hanging out in a petri dish and the number is doubling every three hours. Find an algebraic model for this situation. How many bacteria are there after t hours? When will there be 50,000 bacteria in the dish?

4) The population of a small town in Wisconsin in the year 1890 was 6250. Assume that the population increased at the rate of 5% per year. What was the population in 1915? What was the population in 1940? What is the population today? What limitations on the exponential model does this problem expose?

5) The half-life of a radioactive substance is the amount of time required for half of an amount of the substance to decay. If the half-life of kryptonite is 35 days, how much of an initial 18-gram sample will remain after t days? Draw a graph of this situation, and compute when there will be 2 grams remaining.

6) A ball is dropped from a height of 100 feet on a far off planet. It hits the surface after 4 seconds. Find an algebraic model for the height of the ball as a function of time.

7) Here are the data for growth in three rabbit colonies given in years. In each case...
    a) Sketch a graph of the population as a function of time;
    b) Find the initial population;
    c) Find a reasonable algebraic model for the growth; and
    d) Explain what is happening to each of these colonies in the long run.

Colony A		Colony B		Colony C	
Year	Pop.	Year	Pop.	Year	Pop.
5	1256	5	560	5	800
10	1287	10	542	10	900
15	1319	15	524	15	900
20	1353	20	506	20	800
25	1387	25	488	25	600
30	1422	30	470	30	300

8) The water level in a tidal basin over the course of a twelve-hour period is given by the following data.

Time	Depth of the Water
12:00(noon)	10.1 feet
2:00	18.8 feet
4:00	21.8 feet
6:00	16.2 feet
8:00	8.2 feet
10:00	7.0 feet
12:00(midnight)	12.8 feet

a) Does it make sense that water in a tidal pool would behave this way? Explain.
b) Based on this data, predict the times of high tide. Just how high will it get?
c) When is the depth of the water changing the fastest?
d) Find an algebraic model that fits this data.

9) Read problems 3-34, 3-35 and 3-35 in Core Connections Course 3 (Grade 8) of the College Preparatory Mathematics (CPM) curriculum.
- Answer all the questions posed in those tasks.
- Notice that the x-values (inputs) are now listed in order. The authors did this on purpose. Why do you think they did this? How might this help students to begin to notice key features of linear functions and of quadratic functions?

10) Read problems 3-61 and 3-63 in Core Connections Course 3 (Grade 8) of the College Preparatory Mathematics (CPM) curriculum.
- Answer all the questions posed in those tasks.
- Which part(s) of these tasks ask students to think about the concept of the domain of a function?
- Which part(s) of these tasks ask students to think about the distinction between discrete and continous functions?
- Which part(s) of these tasks ask students to think about the distinction between linear and non-linear functions?
- Thinking about O'Callaghan's competencies, in which part(s) of these tasks are students modeling? In which part(s) of these tasks are students interpreting? In which part(s) of these tasks are students translating?

Class Activity 11:   Picturing Functions

Our world is not a calcified relational system but a realm of change, a realm of variable objects depending on each other; functions is a special kind of dependences, that is, between variables which are distinguished as dependent and independent.
H. Freudenthal
Mathematics as an Educational Task

Here is a challenge for you. See how many of these functions you can sketch without using your calculators. You get a star for the correct general shape. You earn another star for correctly labeling the *x*-axis, and a third star for correctly labeling the *y*-axis. Make sure to talk about the roles of the various constants (fixed numbers). Work on each of these *as a group* – don't just split up the problems among group members.

1) $f(x) = (x-3)^2(x+2)(x-1)$

2) $g(x) = 3x^3 - 4x^2 + 7x$

3) $h(x) = {}^-2x^2 + 5x - 8$

4) $k(x) = 3x - 2$

5) $m(x) = \left(\frac{1}{2}\right)^x + 4$

6) $p(x) = 2\sin(x)$

7) $q(x) = \sin(x) + 3$

8) $r(x) = \cos(2x)$

Which of the above are periodic functions?

Which of the above grow without bound?

Which of the above are **polynomial** functions? *Look up the definition in the glossary.*

What things do you notice about the graphs of polynomials?

What is the maximum number of times that a polynomial of degree *n* can cross the *x*-axis?

What is the minimum number of times that a polynomial of degree *n* must cross the *x*-axis?

Read and Study:

The soul never thinks without a picture.

Aristotle

Let's talk polynomials. These are a special class of functions. Every **polynomial** can be placed in this general form:

$$f(x) = a_n x^n + a_{(n-1)} x^{(n-1)} + a_{(n-2)} x^{(n-2)} + \ldots + a_1 x + a_0.$$

where 'n' is some whole number; it is called the **degree** of the polynomial. All the 'a's in the above expression are real numbers, called the **coefficients**, and the subscripts are just part of their names to label them and to show that they may all be different numbers. We require that a_n not be zero. *Why?*

For example, $g(x) = 2x + 8x^3$ is a polynomial of degree 3. *Decide whether each of the following fits the definition of a polynomial, and if so, state its degree:*

1) $h(x) = x^{\frac{1}{2}} + 3x^2 + 5$

2) $j(x) = 0.34x^9 - 546x^3 + 5$

3) $k(x) = 9(5 - x)(x + 7)(x - 3)^2$

4) $m(x) = \frac{x^4 - 7x}{x}$

Okay, so only the second and the third above qualify as polynomials. The second has degree 9 and the third has degree 4 (*multiply it out and see*). The first function has a power that is not a whole number so it cannot be a polynomial. The last function is almost a polynomial – we know, you think it can be put in the above general form. It can; but as written, it has a domain that does not include zero. So if you wrote it in this form: $m(x) = x^3 - 7$, you would have to note that $x \neq 0$. It inherits this restriction from its original form. But polynomials have unrestricted domains; they can take in all real numbers, so the fourth is disqualified.

Polynomials have lots of great properties. A lot of them stem from the **Fundamental Theorem of Algebra**. It guarantees that every *n*-degree polynomial

$$p(x) = a_n x^n + a_{(n-1)} x^{(n-1)} + a_{(n-2)} x^{(n-2)} + \ldots + a_1 x + a_0$$

has exactly *n* roots (counting double roots as two, triples roots as three, etc.) if we allow complex numbers as roots. Furthermore, if we label the roots $r_1, r_2, \ldots, r_n$, then the polynomial can be written as a product of linear factors as follows:

$$p(x) = a_n (x - r_1)(x - r_2)(x - r_3) \ldots (x - r_n).$$

Another great property of polynomials is that they are continuous. We will define continuity later, but informally this means that the graph of a polynomial will not have any holes, breaks or vertical asymptotes.

Polynomials in completely factored form, like number three above, are fun to graph. We'll talk you through that one so you can see how we think about it. Let

$$k(x) = 9(5 - x)(x + 7)(x - 3)^2.$$

The roots of this polynomial are x = 5, x = -7 (those are both single roots) and x = 3 (a double root). So we can begin by plotting these roots on the graph. This is where the function is zero.

We can also easily check whether the value for the function will be positive or negative between each of these roots, by checking test cases in those intervals.

We can now sketch the polynomial by making a continuous curve through these roots that is positive (above the x-axis) and negative (below the x-axis) on the proper intervals between the roots. *Try this. Sketch a rough graph of the polynomial $k(x) = 9(5 - x)(x + 7)(x - 3)^2$ that shows where the function is positive, negative and zero. Do not worry about the scale on the y axis. Don't use a graphing calculator either. Our goal here is to see what we can figure out about the behavior of a polynomial function by analyzing its formula.*

Look at your graph and pay attention to the behavior of the graph near the roots. The factored form of the polynomial also tells you *how* the graph of the function hits those roots. At x = 5 and x = -7, these are single roots, and the graph goes through the x-axis (like a straight line would); at x = 3, which is a double root, the graph will bounce off the x-axis (like the vertex of a parabola would).

A polynomial that is in completely factored form is pretty special. In general, it is tough to factor a polynomial completely. For one thing, not all polynomials factor unless you allow complex numbers. Take the function $f(x) = x^2 + 4$. It is a polynomial of degree 2 and it seems like it should have two real roots (two places where it crosses the x- axis), but is has none. *Graph it and see.* Its roots involve the number *i* (the square root of negative 1) and so they are both complex roots. *Use the quadratic formula to solve for x and see for yourself.*

For another thing, even if a polynomial had only real roots, we do not have too many tools for factoring them. Degree 1, 2, 3 and 4 polynomials are the exceptions. There are formulas for factoring these. (In the case of degree 2 polynomials, that formula is called the quadratic formula.) But get this: mathematicians have *proved* that no such formulas exist for general polynomials with degrees higher than four. It isn't that we haven't found them yet. They cannot exist.

In the examples on the previous page, we decided that the fourth function was not a polynomial. However, it is the ratio of two polynomials. A **rational function** is a function that is the ratio of two polynomials. For example, $f(x) = \frac{3x^2-1}{x^2+5x+6}$ and $g(x) = \frac{x^5}{x^2+1}$ are rational functions. So is $h(x) = \frac{1}{x}$.

The behavior of a rational function is determined in large part by the behavior of the polynomials that make up the numerator and denominator. In particular, when the denominator is zero, the rational function will be undefined. That means that any values for x that is a root of the denominator polynomial will not be in the domain of the rational function. For example, $f(x) = \frac{3x^2-1}{x^2+5x+6}$ is not defined when x = -3 or x = -2, so these two values are not in the domain of this function. *Which values of x are not in the domain of $g(x) = \frac{x^5}{x^2+1}$? Which values of x are not in the domain of $h(x) = \frac{1}{x}$?*

Homework:

A problem is a chance to do your best.

Duke Ellington

1) Do all the italicized things in the *Read and Study* section.

2) Decide whether each of the following is True or False. In each case, justify your answer.
    a) All third degree polynomials have at least one real root.
    b) All fourth degree polynomials have at least one real root.
    c) There are no sixth degree polynomials that can be factored completely.
    d) A third degree polynomial must have exactly three real roots.
    e) If (x – 3) is a factor of a polynomial, then x = 3 is a root.
    f) Complex roots always come in pairs.

3) Sketch, *without a calculator*, graphs of the following functions of x. Don't worry about the scale on the y axis, but show the location of all roots and the general behavior of the function between the roots.
    a) $y = (3 - x)^3 x^2 (x + 5)^2 (x - 7)$
    b) $y = x^3(x - 5)(x + 4)$
    c) $y = \frac{1}{2}(x + 3)^2 - 6$

4) Determine the domain for the following functions. Explain your reasoning.
    a) $f(x) = 2x(x - 3)^2$
    b) $g(x) = \frac{x+5}{x^2+x}$

5) Build Your Own:

   a) Equation with exactly two solutions $x = 5$, and $x = -7$.

   b) Quadratic Function with roots $x = \frac{5+\sqrt{3}}{2}$ and $x = \frac{5-\sqrt{3}}{2}$

   c) Linear function with exactly one root $x = 6$.

   d) Quadratic function with exactly one root $x = 6$.

   e) An equation with exactly 3 solutions: $x = 0$, $x = 3$, $x = 6$.

   f) Polynomial with roots at exactly $x = -6, x = 2/3, x = \sqrt{2}, x = -\sqrt{2}$

   g) Degree 5 polynomial with roots only at 4 and 7.

6) The graph of a function $y = f(x)$ is sketched below. (The y axis is not shown to scale). Write an appropriate algebraic formula for the function.

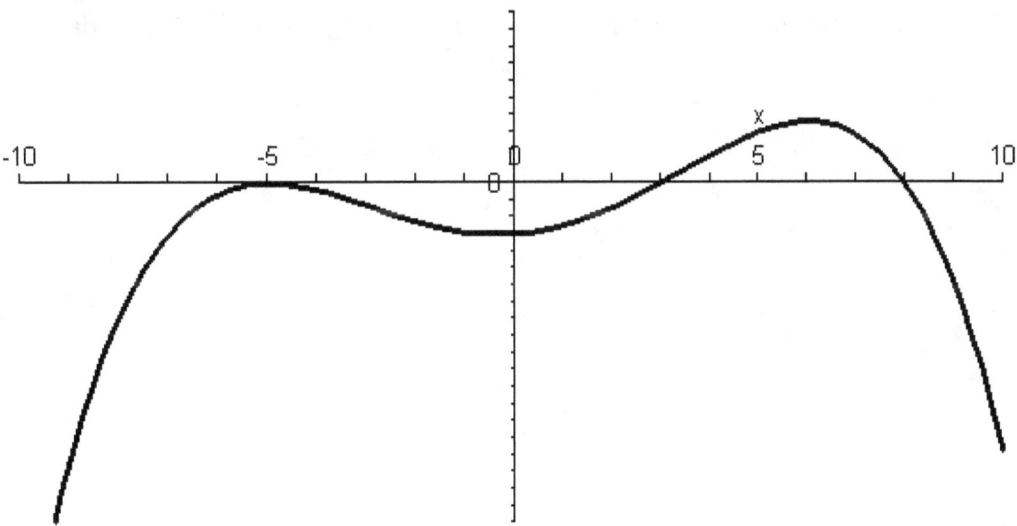

7) Sketch a graph of each of the following functions (You may use your calculator or graphing app). This is your library of basic function shapes.

- $y = f(x) = x$
- $y = f(x) = x^2$
- $y = f(x) = x^3$
- $y = f(x) = 2^x$
- $y = f(x) = \sin(x)$.

a) For each function above, also sketch the graph of $y = -f(x)$
b) For each function above, also sketch the graph of $y = f(-x)$
c) For each function above, also sketch the graph of $y = f(2x)$
d) For each function above, also sketch the graph of $y = 3f(x)$
e) For each function above, also sketch the graph of $y = f(x + 4)$
f) For each function above, also sketch the graph of $y = f(x) + 5$

Summarize the effect that each transformation has on the graph of the function $f(x)$.

8) Do problem 3-66 in Core Connections Course 3 (Grade 8) of the College Preparatory Mathematics (CPM) curriculum. Answer all the questions posed in this task. In what way in this task asking students to think about transformations of graphs? Which of O'Callaghan's competencies does transforming graphs best fit under?

Class Activity 12: Take it to the Limit

Take it to the limit, take it to the limit, take it to the limit, one more time.
The Eagles

Let's wrestle with an important mathematical definition. Suppose $f(x)$ is a function defined on an interval around a point $x = c$ (but not necessarily at c itself).

> The **limit** of the function $f(x)$ as x approaches c is the number L if $f(x)$ is arbitrarily close to L for all x sufficiently close to, but not equal to, c.
> If the limit L exists, we write $\lim_{x \to c} f(x) = L$.

Okay. The first thing a mathematician does when faced with a new definition is to start making up some examples (and non-examples) that help illustrate the idea. So we need a function defined around a point.

Here's a function: $f(x) = \frac{2x^2 - 18}{x - 3}$. Hmm, let's look at the point $x = 3$ because it is interesting. To make this more concrete, complete the table below. Note that $f(x)$ is not defined at $x = 3$. Why? But we don't care about that: limits are all about what happens arbitrarily *close* to the point, not at the point.

x	f(x)
2	
2.5	
2.9	
2.99	
2.999	
3	
3.001	
3.01	
3.1	
3.5	
4	

As x gets close to 3, $f(x)$ gets close to what number L?

How close to 3 does x need to be to have $f(x)$ within 0.001 of L?

How close to 3 does x need to be to have $f(x)$ within 0.00001 of L?

83

How close to 3 does x need to be to have $f(x)$ within ε of L?

(The symbol ε is the Greek letter epsilon, which mathematicians like to use to represent an arbitrarily small number close to zero.)

Notice that for *any* tolerance around L that we give you, you can always find an interval around 3 that puts $f(x)$ within that tolerance. *Compare this to the definition above. Make sure you understand this!* Explain it to someone else in your group and also listen carefully to his or her explanation.

Read and Study:

Quantities, and the ratios of quantities, which in any finite time converge continually to equality, and before the end of that time approach nearer to each other than by any given difference, become ultimately equal.
                                                                    Sir Isaac Newton in Principia

While the definition of limit is static (it doesn't involve motion), when mathematicians are finding limits we often think of finding a limit as involving motion. Suppose *f(x)* is a function defined on an interval around a point *x = c* (but not necessarily at *c* itself).

Motion concept of limit: L is the limit of $f(x)$ as x approaches c if as x gets closer and closer to (but doesn't reach) c, $f(x)$ gets closer and closer to L. And this has to be true when c is approached from either side.

Definition of limit: L is the limit of $f(x)$ as x approaches c if $f(x)$ is arbitrarily close to L for all x sufficiently close to c (but not equal to c).

Spend a few minutes comparing the two definitions. Do they say the same thing? Can you see why one is static and the other involves thinking about motion? We note that it is certainly possible for a limit to 'not exist.' *Try to draw some pictures of functions where the limit does not exist at a point c.* Make sure to talk about this in class.

The first definition is helpful for an intuitive understanding of limit, but it is no good for *proving* things, and you know how we mathematicians love to do that. The second definition is great for doing proofs – and that is why we have settled on it as our official definition. But if you want to think about motion when you do limits, you go right ahead.

Two things that the limit is *not*:

> The limit is *not* a boundary that cannot be crossed (even though that is the way we use it in everyday speech).

> The limit is *not* unreachable. Most limits are actually values taken on by the function. For example, if you have the boring constant function $f(x) = 5$ then the limit as x approaches 2 would be 5. And the function achieves that value all the time (literally). In fact, whenever a limit at a point is the same as the value take by the function at that point, we say the function is continuous there. More on that later.

So why do we care about limits? The answer is there are lots of situations when a function is not defined at a point, but we are still interested in what the function's behavior is like as you approach that point. Every major concept of calculus – derivative, continuity, integral, and the idea of convergence – is defined in terms of limits. *Limit* is the most fundamental concept of calculus; in fact, the idea of *limit* is what distinguishes calculus from algebra, geometry, and the rest of mathematics.

Therefore, in terms of the orderly and logical development of calculus, limits must come first. But the weird thing is that the historical record is just the reverse. For many centuries, the notions of limit were confused and vague, and sometimes mathematicians talked about "infinitely large numbers" and "infinitely small numbers" (this would not be standard talk today). The term *limit* in our modern sense is a product of the late 18th and early 19th centuries and our modern definition is less than 150 years old. Up until this time, there were only rare instances in which the idea of the limit was used rigorously.

To illustrate how we can use the definition to prove a limit exists, let's consider the function $f(x) = 3x - 1$. Yes, we know this is not a very complicated function, and hopefully you really understand this function very well. (That makes it a perfect example to use to try to understand a new concept like the definition of a limit.) We know it is a linear function, and that as x changes by 1, the y value will change by 3. We know the graph is a straight line with slope 3, and a y-intercept at $y = -1$. That's all very good.

Now let's investigate the limit of this function as x approaches 2. In our new notation, we want to evaluate $\lim_{x \to 2}(3x - 1)$. Yes, we know that $f(2) = 5$. But this is not what we are getting at. The idea of the limit is to ignore whatever we may know about what the function is like **at** $x = 2$, and instead just look at what happens **near** $x = 2$. Ok, so ignoring what happens **at** $x = 2$, what happens as x gets closer and closer to 2? (This is the motion concept of the limit). From what we know about the behavior of linear functions and straight lines, we know that the y value (also called the *function value*) will get closer and closer to 5. Great! So we have a conjecture, namely that $\lim_{x \to 2}(3x - 1) = 5$.

But to prove that this is indeed the limit, we must show that this satisfies the definition of the limit. First of all, according to the definition $\lim_{x \to 2}(3x - 1) = 5$ would mean that we can make $f(x) = 3x - 1$ arbitrarily close to 5 by making x sufficiently close (but not equal to) 2. But what is meant by *arbitrarily* close and *sufficiently* close?

Arbitrarily close means as close as you or anyone could want. For example, suppose you wanted $f(x) = 3x - 1$ to be within a half a unit of 5. That means you want $3x - 1$ to be between 4.5 and 5.5. Then how close does x need to be to 2 to make this happen? In other words, how close to 2 is sufficient? With a little arithmetic, you will find that whenever x is within one sixth of 2, then $3x - 1$ will be within a half of 5. *Do this arithmetic now.*

But what if someone wanted you to get closer? Suppose you needed to get $f(x) = 3x - 1$ to be within one-hundredth of 5. Then how close does x need to be to 2? Again, we can oblige, and find that as long as x is within 1/300 of 2, that will be within 1/100 of 2.

Now we can prove that $\lim_{x\to 2}(3x - 1) = 5$ is true once and for all. Suppose someone required $f(x) = 3x - 1$ to be within ε of 5. (Recall this is a lower-case epsilon, a Greek letter used by mathematicians to denote an arbitrarily small quantity). Then how close does x need to be to 2? We can answer this with a little algebra. $3x - 1$ can be as high as $5 + \varepsilon$, and as low as $5 - \varepsilon$. So let's solve the following inequality:

$$5 - \varepsilon < 3x - 1 < 5 + \varepsilon$$

$$6 - \frac{\varepsilon}{3} < 3x < 6 + \frac{\varepsilon}{3}$$

$$2 - \frac{\varepsilon}{3} < x < 2 + \frac{\varepsilon}{3}$$

So that means x needs to be within $\frac{\varepsilon}{3}$ of 2. Here's the big idea: no matter how close someone wants $f(x) = 3x - 1$ to be to 5, we can find how close x needs to be to 2 to make this happen. We can say that $f(x) = 3x - 1$ will be within ε of 5, whenever x is within $\frac{\varepsilon}{3}$ of 2. That's what it means for $\lim_{x\to 2}(3x - 1)$ to equal 5.

In our last example, for the function $f(x) = 3x - 1$, we saw that the $\lim_{x\to 2}(3x - 1) = 5$. We also know that $f(2) = 5$. That is, the limit is actually the same as the function value at that point. When this happens, we say that the function is **continuous** at that point.

Formally, a function $f(x)$ is <u>**continuous**</u> at x = c means that $\lim_{x\to c} f(x) = f(c)$.

We say that a function is <u>**continuous on an interval**</u> if it is continuous at every point in that interval. In general, the common functions we will be dealing with tend to be continuous over most of their domain. Polynomial functions we know are continuous on the entire real line. Rational functions are continuous except at their vertical asymptotes. It is at the points where the function is not continuous that things get really interesting.

For example, the function we studied in the class activity was $f(x) = \frac{2x^2 - 18}{x - 3}$. We saw that as x approaches 3, the limit of this function is 12. However, the function is not defined at x = 3. So this function is not continuous at x = 3.

If we looked at a graph of this function, we would see that it looks exactly like the graph of $y = 2x + 6$, except for a hole at x = 3. We call this type of discontinuity a <u>**removable discontinuity**</u>. (The reason it is called a removable discontinuity is that if we wanted to make the function continuous at x = 3, it would be very easy to do so. We would simply define the function to be 12 at x = 3, which would be equivalent to replacing the function by the formula $f(x) = 2x + 6$.)

For another example, consider $g(x) = \frac{1}{x-3}$. This function is also undefined at x = 3, and therefore cannot be continuous at x = 3. However, in this case, the limit as x approaches 3 does not exist either. To see this, we can do another numerical investigation.

x	f(x)
2.9	-10
2.99	-100
2.999	-1000
2.9999	-10000
2.99999	-100000
3	undefined
3.00001	100000
3.0001	10000
3.001	1000
3.01	100
3.1	10

We see that as x approaches 3 from below, the value of the function grows more and more negative without bound, and as x approaches 3 from above, the value of the function grows more and more positive without bound. That is, the graph of $g(x) = \frac{1}{x-3}$ has a vertical asymptote at x = 3 and we say that $g(x)$ has an **infinite discontinuity** there.

Now consider the function $h(x)$ whose graph is shown below. Unlike the previous two examples, this function is defined at x = 3. The filled in dot at the point (3, 5) indicates that $h(3)$ is defined to be 5. However, the limit as x approaches 3 does not exist. *Why not? Explain why the limit as x approaches 3 is not 5. For that matter, explain why the limit as x approaches 3 is not 1 either. Use the definition of limit to make your argument.* This type of discontinuity is called a **jump discontinuity**.

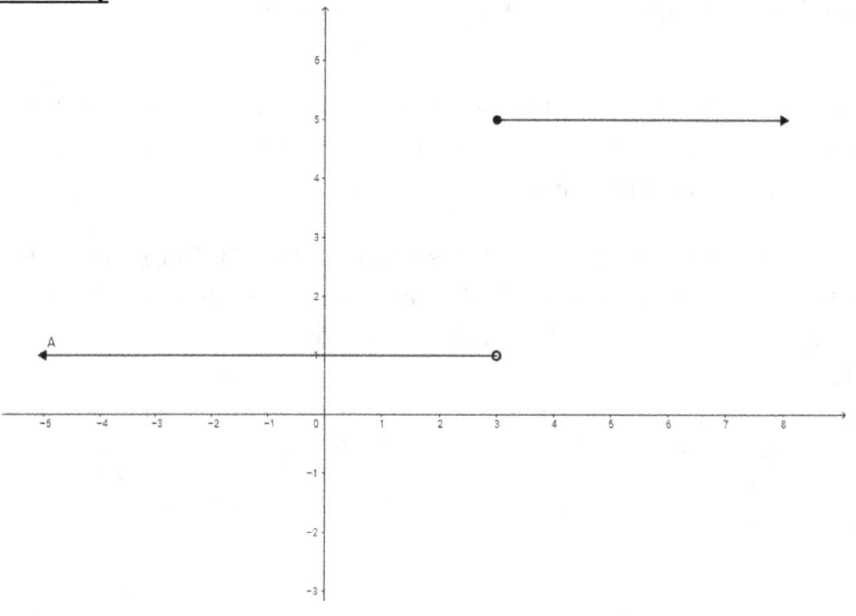

Recall that definition of continuity is that $\lim_{x \to c} f(x) = f(c)$. That is, both the limit and the function value must exist at $x = c$, and they must equal each other. In each type of discontinuity, (removable, infinite and jump), either the function value $f(c)$ does not exist, or the limit $\lim_{x \to c} f(x)$ does not exist, or they both exist but do not equal each other.

Homework:

I can't tell you where Ramanujan got his mathematical desires – I should call them desires, because that's what you really require first. Not so much the ability, but the desire to play around, and to notice. That's what he had to do ... play with numbers until he had played enough to discover things that nobody else knew. He was playing.

Richard Feynman

1) Do all the italicized things in the *Read and Study* section.

2) Use a graphing calculator or spreadsheet to determine the value of the following limits, and illustrate the definition of the limit by finding how close x needs to be to c in order for $f(x)$ to be within $\varepsilon = 0.001$ of L.

    a) $\lim_{x \to 2} (x^3 - 3x + 3)$

    b) $\lim_{x \to 5} \frac{x^2 - 7x + 10}{x^2 - 4x - 5}$

    c) $\lim_{x \to 0} \frac{\sin(x)}{x}$

3) Use the definition of the limit to prove these limits are correct by finding how close x needs to be to c in order for $f(x)$ to be within ε of L. Find a numerical value for this distance when $\varepsilon = 0.01$ and also find a general formula for this distance as a function of ε.

    a) $\lim_{x \to 1} \frac{x^2 - 1}{x - 1} = 2$

    b) $\lim_{x \to 4} (x^2 - 8x + 32) = 16$

    c) $\lim_{x \to 2} \frac{1}{2x + 4} = \frac{1}{8}$

4) Consider the function $k(x) = \frac{x^2-4x+4}{x-2}$. Find the limits $\lim_{x \to 2} k(x)$ and $\lim_{x \to -2} k(x)$. Locate and classify each of the discontinuities for $k(x)$.

5) Now consider the function $g(x) = \frac{x^2-4x+4}{x+2}$. Find the limits $\lim_{x \to 2} g(x)$ and $\lim_{x \to -2} g(x)$. Locate and classify each of the discontinuities for $g(x)$.

6) The graph of a function f(x) is shown below.

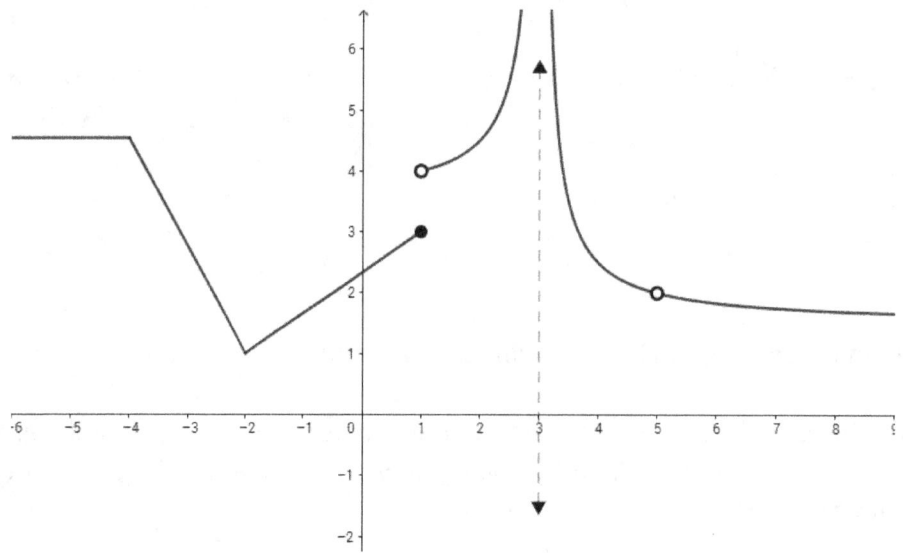

Find each of the following limits, then locate and classify each of the discontinuities.

a) $\lim_{x \to 5} f(x)$
b) $\lim_{x \to -2} f(x)$
c) $\lim_{x \to 1} f(x)$
d) $\lim_{x \to 3} f(x)$

7) Sketch the graph of a function with the following properties:
   a) $f(5) = 7$ and $\lim_{x \to 5} f(x) = -3$
   b) $f(5) = 7$ and $\lim_{x \to 5} f(x)$ does not exist.
   c) $f(5)$ is undefined and $\lim_{x \to 5} f(x)$ does not exist.

8) Determine whether each of the following are true or false and explain your reasoning:
   a) If $\lim_{x \to c} f(x)$ exists, then $f(x)$ is continuous at c.
   b) If $f(c)$ exists, the $f(x)$ is continuous at c.
   c) If $f(x)$ is continuous at c, then $f(c)$ exists.
   d) If $f(x)$ is continuous at c, then $\lim_{x \to c} f(x)$ exists.
   e) $f(x)$ is continuous at c if and only if $\lim_{x \to c} f(x) = f(c)$.

9) Read problems 8-121 and 8-122 in Core Connections Course 3 (Grade 8) of the College Preparatory Mathematics (CPM) curriculum. Make the graphs for the functions in 8-121

and for each graph answer all the questions (both italicized and not) that are posed in 8-122.

Class Activity 13:   Growing All the Time

Strength and growth come only through continuous effort and struggle.
Napoleon Hill

Suppose you have one unit of something. Let's say you have 1 liter of water. Wait, that doesn't seem very valuable. Ok, how about 1 pound of gold.

If you had 1 pound of gold, and if somehow you were able to increase the amount you had by 100%, then your amount would double to 2 pounds, right?

What if instead, starting again with 1 pound of gold, you increased the amount you had by 50%, then increased this amount by 50%. In other words, you increase your original amount by 50 % twice. Would you still end of with 2 pounds of gold? Explain.

What if you start with 1, then increase by 25% four times?

What if you start with 1, then increase by 20% five times?

What if you start with 1, then increase by 10% ten times?

What if you start with 1, then increase by 5% twenty times?

What if you start with 1, then increase by 1% one hundred times?

Find a formula for increasing your amount by $(100/n)$% a total of n times. What is the limit of this process as n approaches infinity?

Read and Study:

I want to stay as close to the edge as I can without going over. Out on the edge you see all kinds of things you can't see from the center.

Kurt Vonnegut, Jr.

The limit that you found in the class activity is a very important number which we denote with the letter *e*. That is,

$$e = \lim_{n \to \infty} \left(1 + \frac{1}{n}\right)^n.$$

This number *e* is called the natural exponent base, or Euler's number, named after Leonard Euler, though Jacob Bernoulli was the first to study this limit. This number *e* is really cool. First of all, it is the result of an infinite process. Also, it is irrational. But moreover, when we use this number as the base in an exponential function, we get $y = e^x$, which we learn in the next chapter has some amazing properties.

Why is this number so important? Things tend to grow naturally like this. In nature, things do not tend to grow at discrete time steps. While at a microscopic level, growth may occur suddenly, like a cell dividing in two, on a larger level, growth takes place continuously. We would like to be able to mathematically model this continuous growth.

As an example, let's consider compound interest. Though this is not exactly a natural phenomenon, it shows us how we can model natural phenomena. Suppose you invest $1000 at 8% interest compounded annually. *How much money will you have after one year? After two years? After 10 years? After 50 years?*

An algebraic model for this situation would be $A(t) = 1000(1 + .08)^t$, where $A(t)$ is the amount after *t* years. *Explain why this formula is correct.*

Now, this formula is actually a continuous model. It's not like you have $1000 in your account for a whole year then suddenly after the year is over you suddenly have $1080. So, even though the interest rate is 8% annual interest doesn't mean the bank has to add in your interest just once a year. They can use this model to figure out how much you should have in your account at the end of each month, for example. *According to the model, how much should you have after one month? After 6 months?*

Well, as long as the bank is figuring out your interest every month, let's have them calculate it a new way. Instead of using an 8% interest rate for the year, let's take that 8% and divide it into twelve. So each month, you will be getting 8/12 % interest. This is called 8% annual interest, compounding monthly. *Using this model, how much money will you have after one month? After 6 months? After one year?*

An algebraic model for this situation would be $A(t) = 1000\left(1 + \frac{.08}{12}\right)^{12t}$, where $A(t)$ is the amount after t years. *Explain why this formula is correct.*

Now why stop at compounding only every month? Why not compound it weekly, or daily, or hourly? In general, suppose we invest $1000 at 8% annual interest, and compound the interest n times a year. Then the amount in the account after t years would be $A(t) = 1000\left(1 + \frac{.08}{n}\right)^{nt}$.

Now here's the really neat question: what happens if we compound the interest *continuously*? To do this, we want to let the number of times we compound the interest to approach infinity. That is, we want $A(t) = \lim_{n \to \infty} 1000\left(1 + \frac{.08}{n}\right)^{nt}$. Using our definition of e and a little algebra, this formula becomes $A(t) = 1000e^{.08t}$. In general, the limit of the compound growth formula $A(t) = P\left(1 + \frac{r}{n}\right)^{nt}$ as n goes to infinity is the continuous growth formula $A(t) = Pe^{rt}$.

In the above discussion we found it very useful to talk about the limit of a function as the variable tends to infinity. We say that L is the **limit of f(x) as x approaches infinity** if $f(x)$ is arbitrarily close to L for all x sufficiently large, and write $\lim_{x \to \infty} f(x) = L$.

Similarly, L is the **limit of f(x) as x approaches negative infinity** if $f(x)$ is arbitrarily close to L for all x sufficiently negative, and write $\lim_{x \to -\infty} f(x) = L$.

These limits may not exist. Often, this is because the function does not approach a finite value L, but continues to grow without bound. In this case, we can say the limit does not exist because it is infinite. For example, the function $f(x) = x^2$ grows without bound as x approaches infinity. Thus we can say $\lim_{x \to \infty} x^2$ does not exist since the function is approaching positive infinity. Even though technically the limit does not exist, most mathematicians have no problem writing $\lim_{x \to \infty} x^2 = \infty$ and saying that the limit is (positive) infinity. Also, since as x approaches negative infinity (gets more and more negative without bound), the function value $f(x) = x^2$ grows positively without bound, we can say $\lim_{x \to -\infty} x^2 = \infty$ as well.

We call looking at $\lim_{x \to \infty} f(x)$ and $\lim_{x \to -\infty} f(x)$ looking at the **end behavior** of a function. *Graph the function $f(x) = e^x$ on your calculator.* You will see that as the x-values get large and positive without bound, the function values (y-values) do too. Thus $\lim_{x \to \infty} e^x = \infty$. But as x-values get large and negative without bound, the function values get closer and closer to zero. Thus $\lim_{x \to -\infty} e^x = 0$.

Here is an example involving a more complicated function: $f(x) = \frac{x^2+4x-5}{x-2}$.

We'll walk you through an analysis of some limits for this function. First we'll graph it (below), and then study the graph to make sure we understand the function. Finally, we'll think about its end behavior and discontinuities.

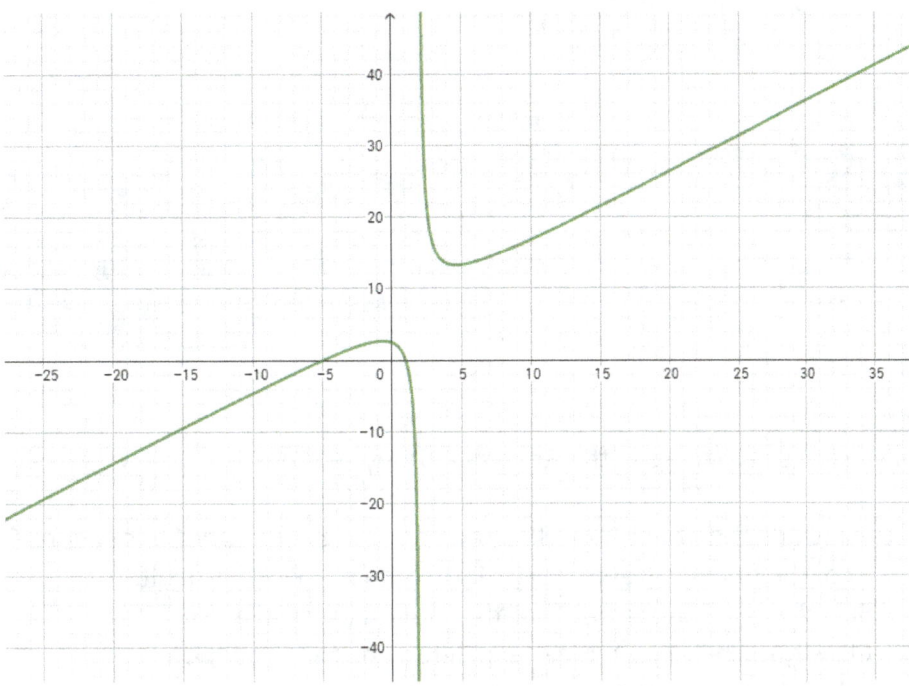

1) We say that this graph has a **vertical asymptote** at x = 2. Why does this happen at the value x = 2? What value does the function have at x = 2? Describe an algebraic method for determining where a function may have a vertical asymptote.
2) Use **polynomial division** to show $f(x) = x + 6 + \frac{7}{x-2}$.
3) We'll call the $x + 6$ part the quotient function $q(x)$ and we'll call the $\frac{7}{x-2}$ part the remainder function $r(x)$.
4) What happens to $r(x)$ as x gets more and more positive? What happens to $q(x)$? to $f(x)$? Do you see that the function behaves like $r(x)$ near 2 and like $q(x)$ away from $x = 2$? What happens to $f(x)$ as x gets more and more negative?

We say that $\lim_{x \to \infty} f(x) = \infty$, and it approaches infinity like the line $y = x - 6$. What is the limit of $f(x)$ as x goes to negative infinity?

We say $\lim_{x \to 2} f(x)$ does not exist because when x is close to 2, $f(x)$ is not close to any specific number. Nor does it approach 'infinity' or 'negative infinity' because as x approaches 2 from the left, the function goes down, while as x approaches 2 from the right, the function goes up.

If however, you have a function that does go off in the same direction on either side of a vertical asymptote, than it is common to say that the limit is infinity, or negative infinity, depending on whether the function is going up or down near the asymptote. For example, consider $g(x) = \frac{1}{x^2}$. *Sketch the graph of this function.* You will see that $g(x)$ has a vertical asymptote at $x = 0$, and the function approaches positive infinity from both sides of the asymptote. So in this case we can say $\lim_{x \to 0} g(x) = \infty$.

Homework:

One thing that never ceases to amaze me, along with the growth of vegetation from the earth and of hair from the head, is the growth of understanding.
Alice Walker

1) Do all the italicized things in the Read and Study section.

2) Read the introduction to section 8.1.1 and do task 8-1 in Core Connections Course 3 (Grade 8) of the College Preparatory Mathematics (CPM) curriculum.
    a. Sketch reasonable graphs for the the functions in a. – d.
    b. Classify each function type (e.g., linear, quadratic, rational, etc).
    c. Which task(s) did we do previously in our textbook that are similar to this one?
    d. Which of O'Callaghan's competencies does this task best address?

3) Read the introduction to section 8.1.2 and do tasks 8-13, 8-14 and 8-15 in Core Connections Course 3 (Grade 8) of the College Preparatory Mathematics (CPM) curriculum.
    a. Answer all the questions posed in these three tasks.
    b. What part(s) of these tasks ask students to compare features of linear and exponential growth.
    c. What part(s) of these tasks ask students to think recursively? What part(s) of these tasks ask students to think functionally?
    d. Thinking about O'Callaghan's competencies, in which part(s) of these tasks are students modeling? In which part(s) of these tasks are students interpreting? In which part(s) of these tasks are students translating?

4) If you invest $1000 at 8% annual interest, how much will you have at the end of 1 year, if the interest is compounded annually? Monthly? Weekly? Daily? Hourly? Continuously? Give answers to the nearest mil (tenth of a cent).

5) Derive the continuous growth formula $A(t) = Pe^{rt}$ from the compound growth formula $A(t) = P\left(1 + \frac{r}{n}\right)^{nt}$ and the definition of e. (Suggestion: try making the substitution $m = \frac{n}{r}$).

6) Read problem 3-105 in Core Connections Course 3 (Grade 8) of the College Preparatory Mathematics (CPM) curriculum.
    a. Answer all the questions posed in this task.
    b. Classify the function that you found to fit the data in the table. (e.g. is it a linear, quadratic, a higher degree polynomial, rational, or exponential function).
    c. Describe any discontinuities and the end behavior of this function.

7) Consider the function $h(x) = \frac{x^3 - 2x^2 - 6x - 4}{x - 3}$.
    a. Predict the end behavior of this function.
    b. Now graph it to check your prediction.
    c. Do the polynomial division and describe how each piece affects the behavior of the function. Which part is dominant for large x? Near x = 3?
    d. Find $\lim_{x \to 3} h(x)$
    e. Find $\lim_{x \to \infty} h(x)$
    f. Find $\lim_{x \to -\infty} h(x)$

8) For each of the functions below, determine the following:
    - The location (x-value) of any discontinuity.
    - The limit of the function as x approaches the discontinuity.
    - The limit of the function as x approaches infinity and negative infinity.
    a. $j(x) = \frac{x^3 - 2x^2 + 6}{x^2 - 4}$
    b. $k(x) = \frac{x^2 - 6x + 9}{x - 3}$
    c. $l(x) = \frac{1}{x^2 - 4x + 4}$
    d. $m(x) = \frac{x^2 - 25}{x^2 + 3x - 10}$

9) Based on your analysis of the rational functions in this section, make some general conjectures about rational functions and their graphs. Specifically, what can you say in general regarding the existence and location of discontinuities, the types of discontinuities that are possible and how you can tell which type a given function have, the kinds of end behaviors that are possible, and how you can tell what type of end behavior a given function will have.

Class Activity 14: Circle Gets the Square

The number π is defined as the ratio of the circumference of a circle to its diameter. That is,

$$\pi = \frac{C}{D},$$

In this activity, you will find lower and upper bounds for the number π by approximating the circumference of a circle with diameter 1. (*Why a diameter of 1?*)

1. First, let's approximate the circumference with perimeters of squares.
    a. Find the perimeter of the inscribed square in a circle with diameter 1.
    b. Find the perimeter of the circumscribed square in a circle with diameter 1.
    c. Use these results to put the value of π between a lower and an upper bound.
    d. Use the average of the upper and lower bound to find an approximation for π.

2. Next, let's approximate the circumference with perimeters of hexagons.
    a. Find the perimeter of the inscribed hexagon in a circle with diameter 1.
    b. Find the perimeter of the circumscribed hexagon in a circle with diameter 1.
    c. Use these results to put the value of π between a lower and an upper bound.
    d. Use the average of the upper and lower bound to find an approximation for π.

3. How could we use perimeters to find even better approximations for the number π? Suppose we wanted to get within 0.001 of the true value of π. Could we do it by calculating perimeters of polygons? Explain.

Read and Study:

Eureka! Eureka.

Archimedes

It has long been known that the ratio of the circumference of a circle to its diameter is a constant, and many ancient cultures, such as the Babylonians, Egyptians, and the Chinese, had come up with approximations for this constant. However, the ancient Greek engineer and mathematician Archimedes (about 200 B.C.) was the first to come up with a method whereby one can get as accurate an approximation as desired. Here is Archimedes' idea (*reductio ad absurdum* meaning 'reduction to absurdity'- an infinite process) for estimating the value of π.

Recall that π is defined as the number of times the diameter (*d*) of a circle fits into its circumference (*C*). The ratio $\frac{C}{d}$ is constant for all circles and this constant is called π.

Now, suppose we start with a regular hexagon inscribed in a circle. We note that the perimeter of the hexagon roughly estimates the circumference of the circle. Suppose we also make a hexagon that circumscribes the circle. This one's perimeter also approximates the circumference. We further note that if we bisect each edge of the hexagons and make a couple regular dodecagons, the perimeters of those figures would be better approximations. Think about it: the perimeter of the inscribed (and circumscribed) regular *n*-gon becomes closer and closer to the circle's circumference as *n* gets big.

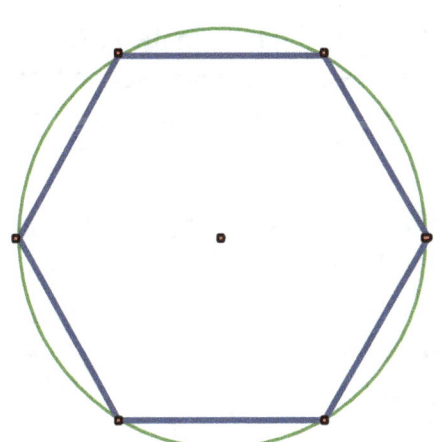 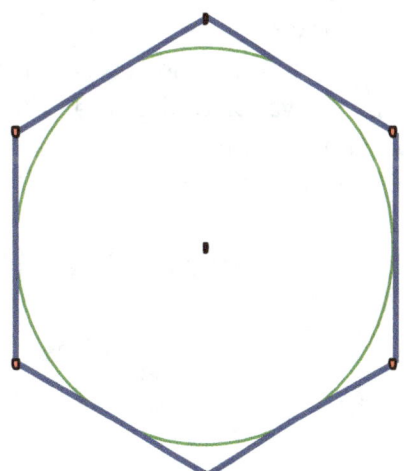

Get this: Archimedes used 96-gons and computed π to be between $3\frac{10}{71}$ and $3\frac{1}{7}$. You should be very impressed; Archimedes had no calculator.

Homework:

Exploring pi is like exploring the universe.
David Chudnovsky

1) The ancient Greeks didn't use decimals, only fractions. Find decimal approximations for those lower and upper bounds for π that Archimedes found using 96-gons.

2) The upper bound that Archimedes found for π, namely 22/7, is still commonly used as an approximation. Use a calculator to determine which value is a closer to the true value of π: 3.14 or 22/7?

3) Find the average of Archimedes' upper and lower bounds. As an approximation to π, this average value is accurate to how many decimal places?

4) A key formula about circles that is taught in middle school is $A = \pi r^2$ for the area of a circle. In section 9.1.2 of Core Connections, Course 2 from the College Preparatory Mathematics (CPM) curriculum, seventh graders are asked to derive this formula.

   a) Do problems 9-23, 9-24, 9-25, 9-26 and 9-27 in that section.
   b) In what way are seventh graders being introduced to the idea of an infinite process in these tasks?
   c) In what way do these tasks demonstrate that the value of π is approximately 3?
   d) In the derivation of the area formula $A = \pi r^2$ in problem 9-26, students use the fact that the circumference of a circle is equal to πd. Explain why the circumference of a circle is equal to πd.

Chapter Three: The Derivative

Class Activity 15:   Walking the Line

Because you're mine, I walk the line.

Johnny Cash

This activity uses technology in the form of TI's CBR (Calculator Based Ranger) and gives you the opportunity to create a graph based on how fast and in what direction you walk. As you work through the problems, keep in mind that you are creating a relationship between two variables. After you are finished, you will have had a hands-on experience with slope and the output intercept.

Before beginning this activity, you will need to link the CBR with your calculator and download the RANGER program. Complete directions are contained in the instruction booklet that comes with the CBR.

Once you and your partner are linked, select the RANGER program from the APPS/CBL-CRB menu. From the main menu of the RANGER program select #3 (Applications) and then choose either meters or feet for units. You will now have a choice of three activities. First choose #1 (Dist Match). In this activity you will attempt to walk (match by walking) a distance graph.

Follow the directions given on the screen. The calculator will show a piecewise linear graph – note that your classmates' screens will vary due to randomly generated graphs. This is okay! Before pressing any keys, examine the graph on your calculator and answer the following questions:

1) Which axis represents time (in seconds)? Is time the independent or dependent variable? How do you know? Which axis represents distance? Is distance the independent or dependent variable? How do you know?

2) Describe how you should walk a path that would look like the one shown. How fast must you walk and in what direction? Explain your reasoning. How is the starting position determined?

Now stand in front of the CBR, press ENTER, then walk the graph shown on your calculator. After you have finished your walk, compare your walk to the initial graph. How well did you match? What do you need to do differently in your walk to have a better match?

Press ENTER and try #1 (Same Match). This choice allows you to try a second walk to match the same distance graph. Repeat your walk and try to get one that is identical to the graph shown on your calculator. Once you have a best match, try another graph by selecting #2 (New Match) and repeat the procedures above. Make sure to take turns with your partner in doing the actual "walking."

After you are successful in matching distance graphs, discuss and answer the following questions:

3) What action represents a change in the slope of the line? Use the works "walking faster," "walking slower," and "stopping" where appropriate.

4) What feature on the graph tells you to turn and walk in the opposite direction?

5) What is the difference between speed and velocity? Which one can be negative? How do we interpret speed and velocity on the "distance from" graph?

6) How do we determine the total distance we have walked?

7) How would we interpret a negative portion of a "distance from" graph?

Now you are ready to challenge yourselves with similar experiments and graphical analysis with time and velocity data using the VEL MATCH application. Return to the applications menu of the RANGER program and choose #2 (Vel Match). Discuss with your partner what kind of walk you need to make in order to match this graph of velocity over time. Again you need to decide where to start, how fast and in what direction to walk to collect data that looks like the given graph. Don't worry if you find it much more difficult to accomplish a good match. Talk with your partner about the possible reasons your walks are not matching. Remedy any of the reasons that have to do with misinterpretations of the original velocity graph and try again until you get a better match.

When you and your partner have successfully matched a velocity walk, discuss and answer the following questions:

8) What action represents a change in the slope of the line on the velocity graph? Use the words "speeding up," "slowing down," "walking at a constant rate," and "stopping" where appropriate.

9) What feature on the graph tells you to turn and walk in the opposite direction?

10) How do we interpret speed and velocity on the velocity graph?

11) How can we determine the total distance we have walked from a velocity graph?

12) How would we interpret a negative portion of a velocity graph?

Class Activity 15B: Toying Around

Play is the highest form of research

Albert Einstein

Record 3 video clips of moving toys. Then for each toy,

- Sketch a plausible graph of the position of the toy as a function of time.

- Sketch plausible graph of the velocity of the toy as a function of time.

Now record position and time data from the video and use a spreadsheet to make a graph of the position of the toy as a function of time.

- How does this graph of the position of the toy compare with your prediction?

Now decide how to use the data we have to approximate the velocity of the toy as a function of time and use the spreadsheet make a graph of these velocities.

- How does this graph of the velocity of the toy compare with your prediction?

Based on your graphs of the position functions and the velocity functions, write your best answers to the following:

1. What is meant by average velocity? Write a definition.

2. What's the difference between speed and velocity? Which can be negative?

3. How can you interpret the speed and velocity of the toy by looking at the position graph.

4. Describe what feature of the position graph corresponds to the toy

    a. moving at a constant rate

    b. slowing down

    c. speeding up

    d. not moving

    e. turning around to move in the opposite direction

5. Describe what feature of the velocity graph corresponds to the toy

    a. moving at a constant rate

    b. slowing down

    c. speeding up

    d. not moving

    e. turning around to move in the opposite direction

6. How can you determine the total distance traveled by the toy by looking at the graph of the position function?

7. How can you determine the total distance traveled by the toy by looking at the graph of the velocity function?

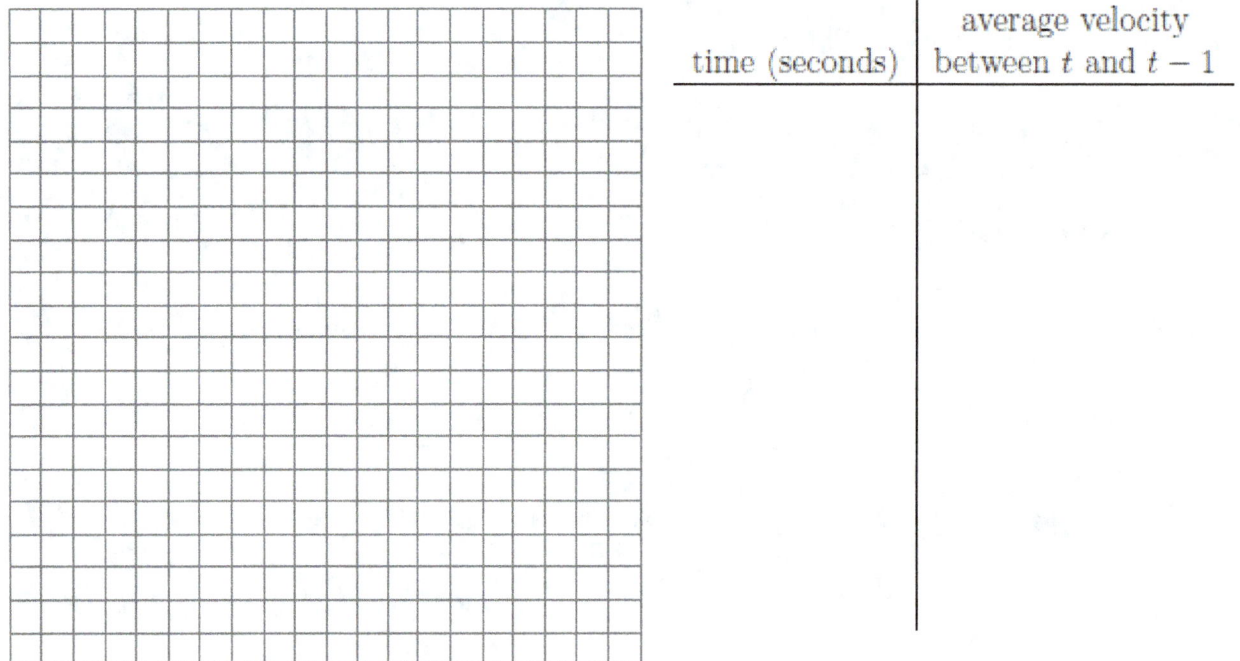

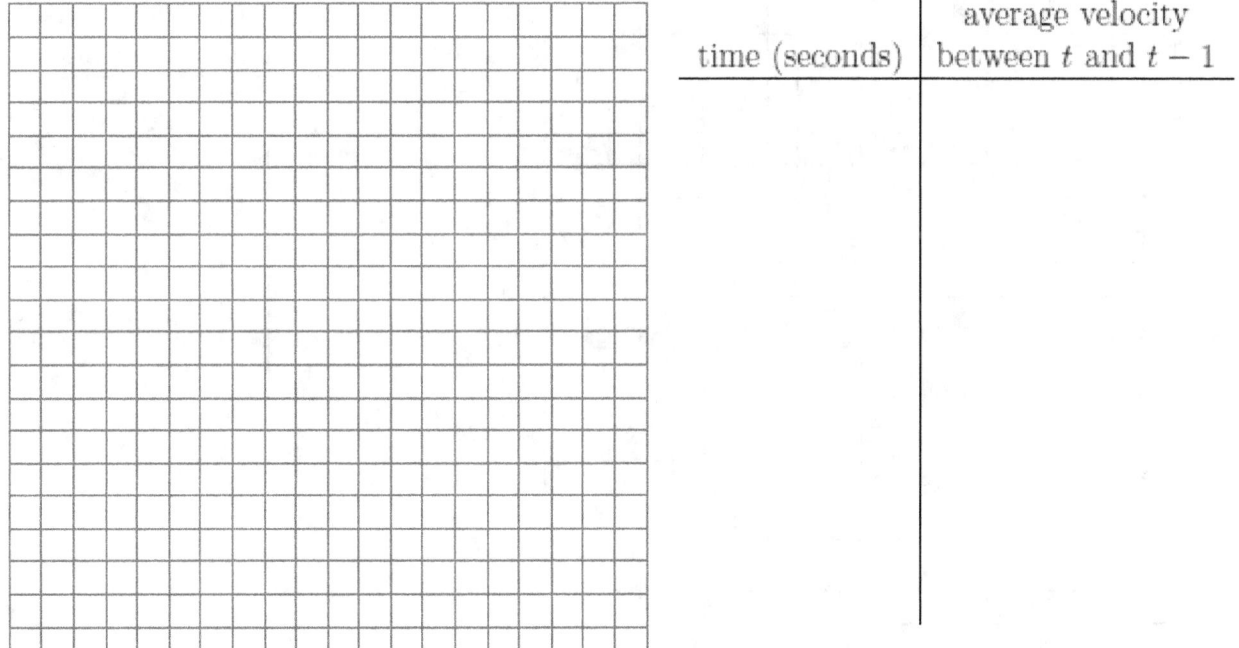

time (seconds)	position

time (seconds)	average velocity between t and $t-1$

Connections to Teaching:

What you have been obligated to discover for yourself leaves a path in your mind which you can use again when the need arises.
G. Polya. Mathematical Discovery: On Understanding,
Learning and Teaching Problem Solving

How is calculus connected to middle grades mathematics? Before we can go there, we probably need to tell you what calculus is all about. Listen up. This is important. **Calculus is the mathematics of change.** When we make a mathematical model to describe some aspect of our world, we usually have a function that relates two variables. We want to know, if we change one thing, how does that affect the other? How fast does one variable change with respect to the other?

An important example of change is motion. Thus *calculus is also the mathematics of motion*. For example, calculus involves techniques for finding the velocity of an object given a function that describes the object's position, and visa versa. Trust us, when analyzing change and motion, limits come up all over the place.

An understanding of calculus begins with experiences measuring and representing motion. As a teacher it is important to have your students do things like:

- walk at precise speeds and collect motion data (Walk down the hall at a speed of 2 feet/second.)
- discuss what objects that might move at a given speed (What might move at 100 feet/second?)
- show position functions (like distance from home on a trip) visually
- measure speeds of objects using a stop watch and measuring tape
- examine connections between 'how far' and 'how fast'
- write equations to model motion
- create time and distance graphs
- attend to units and scale (What if we change to feet/second? miles/hour?)
- interpret position and velocity graphs
- write stories to match given graphs
- compare two position functions or two velocity functions (Which car won the race?)
- solve problems involving distance and speed (How far did the car travel? What was the average speed?)
- examine other rates of change (temperature change, water level, growth)

Class Activity 16:   Baseball Velocity

The source of all great mathematics is the special case, the concrete example.
Paul Halmos, I want to be a Mathematician

Walter Johnson, one of the greatest pitchers of all time, throws a ball straight up from the ground. It reaches a maximum height of 144 feet after 3 seconds.

> Sketch a reasonable graph of the height (label it *h*) of the ball as a function of time (labeled *t*). Is this the path the ball follows? Explain.
>
> Find an algebraic model for the height of the ball as a function of time.

Position and velocity are signed quantities, i.e., they can be positive or negative. Let's call "up" the positive direction in this problem. We also want to distinguish between the **instantaneous velocity** of the ball and the **average velocity** of the ball over a time interval.

1) Decide what is meant by 'average velocity' over a time interval and write a definition. How can we compute the average velocity over the time interval, say [0, 3]? Over the time interval [2, 4]? How does this value (of the average velocity) relate to the graph of the height function?

2) Decide what is meant by 'instantaneous velocity' at a given time and write a definition. In this situation, when is the instantaneous velocity of the ball positive? Zero? Negative? Draw a rough sketch of a graph of instantaneous velocity as a function of time for this situation. How does the instantaneous velocity relate to the graph of the height function?

3) Find the average velocity of the ball over each of the following intervals.

Time Interval	Average Velocity
[0,6]	
[0,3]	
[0,1]	
[0,0.1]	
[0,0.01]	
[0,0.001]	

4) What is the velocity of the ball at the instant that Johnson threw it? (We call this the initial velocity and often denote it v_0.)

5) Find the instantaneous velocity of the ball at one second. At two seconds. At four seconds. Do the velocities you calculated make sense in the context of the problem? Explain.

Read and Study:

I'm very good at integral and differential calculus, I know the scientific names of animalculous; In short, in matters vegetable, animal and mineral, I am the very model of a modern Major-General.

W.S. Gilbert
The Pirates of Penzance

The **derivative** of a function at a point measures the *instantaneous rate of change of that function at that point*. Graphically, the derivative of a function at a point is *the slope of the tangent to the curve at that point*. We just said a lot. *Read it again*. Let's have a look at this from the geometric standpoint. We start by thinking about how to find the average rate of change of a function on the interval $[c, c + \Delta x]$. (The 'delta' (Δ) is typical mathematical notation to indicate a change, so when a mathematician sees Δx, she thinks 'change in x' from the notation alone.)

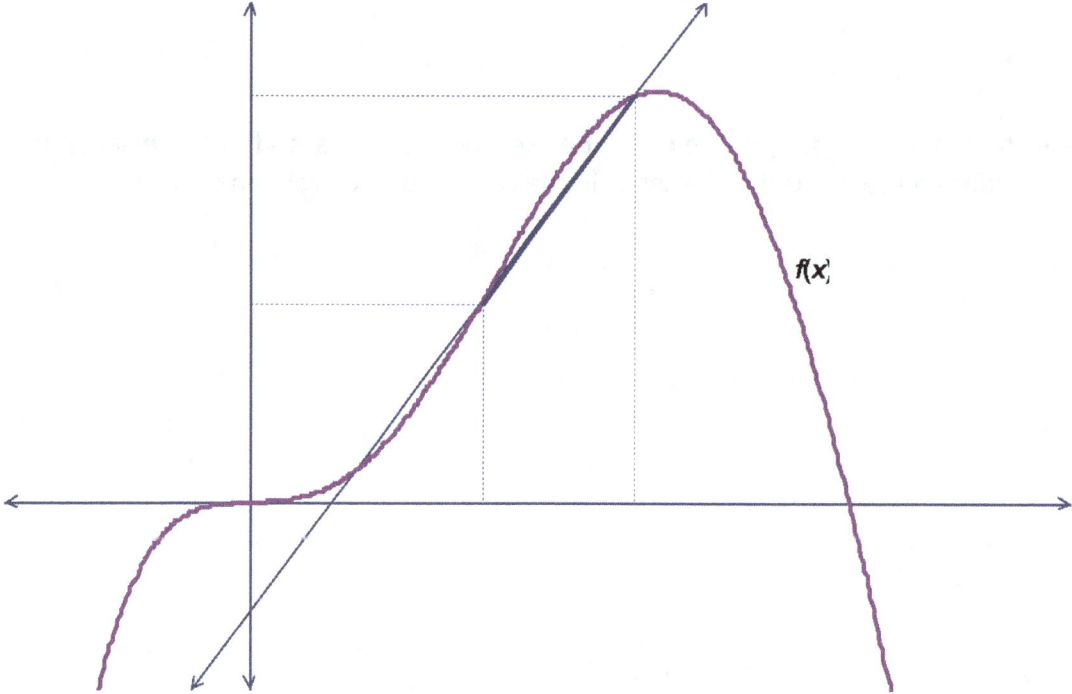

So the function $f(x)$ shown above changes from a value of $f(c)$ to a value of $f(c + \Delta x)$ as we move a length Δx from c. If f were a distance function, like Johnson's throw - it would be a pretty weird one - but anyway, then the ball would have traveled from a height of $f(c)$ to a height of $f(c + \Delta x)$ in Δx seconds (from time c to time $c + \Delta x$). *Think about this and make sure you follow it before you go any further.* So **the average rate of change** of the function from c to $c + \Delta x$ would be given by the following formula:

$$\frac{f(c + \Delta x) - f(c)}{\Delta x}$$

In the case of a throw, the above formula would be change in the height of the ball over the time interval from c seconds to (c + Δx) seconds, divided by that length of time. In other words, the average velocity during that time period. *Notice that geometrically this is the slope of the secant line shown on the above graph. Have a look.*

How close this is to the *instantaneous velocity* at exactly time c depends on how small Δx is. If it's super small, then the average velocity will be very close to the instantaneous velocity. But we're doing calculus here. We're not afraid to go super duper small. We're not even afraid to think about the *limit* of the average velocity as Δx approaches zero. In fact, that's how we *define* the instantaneous velocity at c. And we call this the *derivative* of the function at c.

Suppose f(x) is a continuous function over an interval a < x < b. The **derivative of f(x) at the point c** in that interval, which we denote $f'(c)$, is given by the following limit (if it exists):

$$f'(c) = \lim_{\Delta x \to 0} \frac{f(c+\Delta x)-f(c)}{\Delta x}$$

This is a BIG idea. *Read it again, then practice explaining it to someone, complete with a helpful picture.*

Another way of thinking about derivatives is like this: Lots of functions look like lines when you zoom in tight on the graph. Take, for example, $g(x) = sin(x)$ at $\frac{\pi}{4}$. *Graph this on your calculator (make sure it is in radian mode) and zoom in on the graph above $x = \frac{\pi}{4}$. Zoom in more. More. More, darn it.* Close in enough, even the sine function looks linear. The slope of that line is the derivative at the point $x = \frac{\pi}{4}$. *What is that slope?*

Now let's take this definition of the derivative out for a spin.

A kid throws a ball straight up. It reaches its maximum height of sixteen feet after 1 second. 'Throwing scenarios' fit a quadratic model. In this case, the height of the ball is captured by the function h(t) = -16t(t – 2) (because it will take the ball the same amount of time to fall back down from it's max height into the kid's hand – because its still being acted on by the same force of gravity) or h(t) = -16t² + 32t. We'll compute the instantaneous velocity at, heck we'll just do it at any time t.

113

$$h'(t) = \lim_{\Delta t \to 0} \frac{h(t+\Delta t) - h(t)}{\Delta t}$$

$$= \lim_{\Delta t \to 0} \frac{(-16(t+\Delta t)^2 + 32(t+\Delta t)) - (-16t^2 + 32t)}{\Delta t}$$

$$= \lim_{\Delta t \to 0} \frac{(-16(t^2 + 2t\Delta t + \Delta t^2) + 32(t+\Delta t)) - (-16t^2 + 32t)}{\Delta t}$$

$$= \lim_{\Delta t \to 0} \frac{-16t^2 - 32t\Delta t - 16\Delta t^2 + 32t + 32\Delta t + 16t^2 + 32t}{\Delta t}$$

$$= \lim_{\Delta t \to 0} \frac{-32t\Delta t - 16\Delta t^2 + 32\Delta t}{\Delta t} \quad \text{(All we're doing is algebra so far.)}$$

$$= \lim_{\Delta t \to 0} \frac{(-32t - 16\Delta t + 32)\Delta t}{\Delta t} \quad \text{(Still doing algebra)}$$

$$= \lim_{\Delta t \to 0} (-32t - 16\Delta t + 32) \quad \text{(We are finally going to take the limit now!)}$$

$$= -32t + 32$$

Okay, what have we done? What does this mean? What we have done is use the definition of the derivative to compute the instantaneous rate of change of the height function at any time *t*. In other words, we have the *velocity function*. Its units are in feet/second. Remember we started with a change in height divided by a change in time? Want to know how fast the ball was going at *t* = 1? Put it in there and see.

$$v(t) = h'(t) = -32t + 32$$

This function is valid from the time the ball is released until just before it is caught.

v(1) = 0 ft/sec. (At time *t* = 1, the ball wasn't moving. Makes sense right? It had reached its maximum height and was about to start falling.)

v(.5) = 16 ft/sec. (At time .5 seconds, the ball was going up at 16 ft/sec.)

v(1.5) = -16 ft/sec. (At time 1.5 seconds, the ball was falling at 16 ft/sec.)

Isn't this just the best? We started with a scenario. We made a model (a function that captured the situation). We found the derivative of that function to get another function that shows the rate of change of the original function. (Hey, and if we found the derivative of the velocity function, we'd have the 'rate of change of velocity' function; that's the acceleration function.)

Don't worry if this doesn't sink in the first couple times you read it. It takes work to understand the derivative.

Connections to Teaching:

> *If I were to describe one process in the training of men which is fundamental to success in any direction, it would be a thoroughgoing training in the habit of accurate observation.*
>
> Eugene G. Grace

Middle grades students can investigate patterns in graphs by observing the relationship between distance and time in the context of someone's "travel" around the neighborhood. Students can begin by describing the walk in terms of 'moving away,' 'moving toward,' 'standing still,' 'moving faster,' and 'moving slower' for increasing, decreasing, zero slope, steeper slope, and less steep slope, respectively. Later they will use the more precise terminology.

For example, consider the graph below of three different animals distance from a given point as they move over time. *How might students answer each of the questions given?*

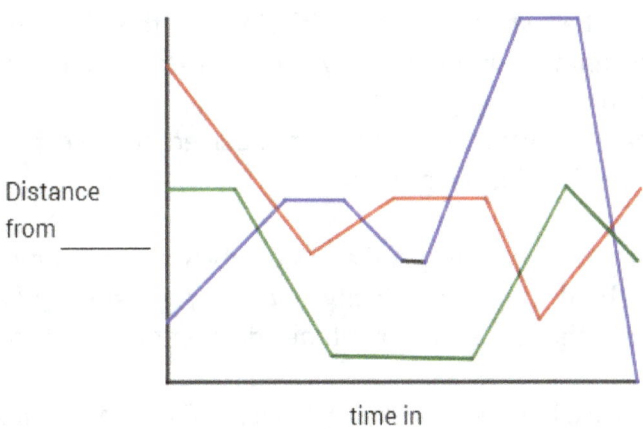

1) Write a story that goes with each graph.
2) What units are you using for time and distance? Why did you pick these units?
3) Draw scales for the distance and time on the graph.
4) For each animal, create a new graph that shows the total distance traveled over time.
5) Which animal moves the fastest in your graph? How do you know?

Now, explain how these questions relate to the idea of the derivative.

Homework:

Vision: the art of seeing things invisible.

Jonathan Swift

1) Do all the italicized things in the *Read and Study* and *Connections* section.

2) In the class activity, you found a model h(t) for a baseball thrown straight up in the air that reached a maximum height of 144 feet after 3 seconds. Use the definition of the derivative to compute the velocity function $v(t) = h'(t)$ for the ball. Sketch a graph of $h(t)$ and a graph of $v(t)$ on the same axis. Compare them.

3) An explosion in the airlock at a Martian Habitation Unit sends the airlock door into the air, reaching a maximum height of 150 feet after 5 seconds.
    a. Draw a graph of the height of the door as a function of time. Assume the door starts at ground level (height of zero), and the moment of the explosion (at time zero).
    b. What is its average velocity during the first 5 seconds of flight? How can this be seen on the graph?
    c. What is its average velocity over the entire flight? How can this be seen on the graph?
    d. Find an algebraic model for the height of the door as a function of time.
    e. How high is the ball when t = 7 seconds? How can this be seen on the graph?
    f. What is its instantaneous velocity at 7 seconds? How can this be seen on the graph? Why is it negative?
    g. What was the initial velocity of the door imparted by the explosion? At what velocity will the door hit the ground?

4) Explain the difference between speed and velocity. Give an example of a situation when speed will always be the same as velocity, and give an example of a situation where the speed is sometimes the same and sometimes different than velocity.

5) Explain the difference between average velocity and instantaneous velocity. How are each of these related to the graph of a position function? Give an example of a situation when the average velocity will be the same as the instantaneous velocity.

6) Use the definition of the derivative to find $f'(x)$ where $f(x) = 4x - 3$.

7) Use the definition of the derivative to find $f'(x)$ where $f(x) = x^2 - x + 7$.

8) The half-life of a certain radioactive substance is 7 years. Suppose we start with 3.5 grams of the substance.
    a) Find a model for the amount of substance (in grams) remaining as a function of time and sketch a careful graph of your function. What is the domain?

b) How many grams remain after 16 years? How does this amount relate to the graph of your function?
c) What is the average rate of decay (in grams/year) of the substance between years 16 and 20? How does this rate relate to the graph of your function?
d) Which rate of decay is faster: the *average* rate of decay from year 16 to year 20 or the *instantaneous* rate of decay at year 16? Explain your answer.
e) Use a graphing calculator to estimate the instantaneous rate of decay (in grams/year) of the substance at year 16? How does this rate relate to the graph of your function?

9) The concept of slope as the rate of change in a function is part of the Common Core State Standards for Grade 8. Read the introduction to section 7.2.2 in the Core Connections Course 3 (grade 8) curriculum and tasks 7-43, 7-44, and 7-45.
   a. Answer all the questions in tasks 7-43, 7-44, and 7-45.
   b. How do these tasks ask students to translate, in the sense of O'Callahan's Competencies?
   c. How are these tasks related to the the definition of the derivative?

10) The population of a herd of deer in Northern Wisconsin is modeled by the function
$$p(t) = 4000 + 500 \sin\left(2\pi t - \frac{\pi}{2}\right)$$
(Here t is measured in years from January 1st, 2000)

   a) How does this population vary with time? Sketch a graph for one year of time. Does this make sense?
   b) Is there a time when the population is at a maximum? How does this time relate to the graph of $p(t)$? A minimum? How does this time relate to the graph of $p(t)$?
   c) At what time of year does the population appear to be growing fastest? How does this time relate to the graph of $p(t)$? When is it decreasing fastest? How does this time relate to the graph of $p(t)$?
   d) Here is a *supposed* model of the derivative function (which measures the rate of change of population over time):
$$p'(t) = 1000\pi \cos\left(2\pi t - \frac{\pi}{2}\right)$$
   What are its units? Sketch a graph of this function for one year of time. Have a look and see if you believe that this is the derivative function just by thinking carefully about the scenario. How is the graph of the derivative function related to the graph of the original function? List any connections that you see.

Class Activity 17: A Derivative Shortcut

Life is good for only two things, discovering mathematics and teaching mathematics.
Siméon Poisson

What is the formula for the derivative of a power function like $f(x) = x^2$? How about the derivative of $g(x) = x^3$? Your tool for exploring these questions is the *definition* of the derivative. Use the definition to actually compute the derivatives of $f(x)$ and $g(x)$ above.

What is your conjecture about the derivative function of $h(x) = x^{25}$? The derivative function for $k(x) = x^n$ for any whole number n? See if you can use the definition of the derivative to argue your conjecture is correct (no fair looking in the *Read and Study* section).

Read and Study:

Mathematics is not a careful march down a well-cleared highway, but a journey into a strange wilderness, where the explorers often get lost. Rigour should be a signal to the historian that the maps have been made, and the real explorers have gone elsewhere.

W. S. Anglin

Before we talk about anything new, let's restate the big ideas about derivatives. We don't want you to lose sight of the forest.

If you have a function $f(x)$, the derivative of that function $f'(x)$ tells you the *instantaneous rate of change of $f(x)$ at each value of x*. So if $f(x)$ tells you say, how deep the water is in a tidal pool, then $f'(x)$ tells you *how fast the depth is changing* as a function of x.

Now let's introduce a new notation for the derivative. Suppose we have a function $y = f(x)$. This gives us two ways of referring to a function, by using $f(x)$ (function notation) and by using the variable y (variable notation). Both are useful. So far we have only denoted derivatives using function notation. Given $y = f(x)$, we defined

$$f'(x) = \lim_{\Delta x \to 0} \frac{f(x+\Delta x) - f(x)}{\Delta x}.$$

Let's rewrite this by using the variable y to refer to the function values. Since the numerator gives the change in the function values, let's just write this as $\Delta y = f(x + \Delta x) - f(x)$. Then the definition looks like this:

$$f'(x) = \lim_{\Delta x \to 0} \frac{\Delta y}{\Delta x}.$$

What Leibniz did was use $\frac{dy}{dx}$ to represent $\lim_{\Delta x \to 0} \frac{\Delta y}{\Delta x}$. Then we have:

$$f'(x) = \frac{dy}{dx}.$$

The notation $\frac{dy}{dx}$ for representing the derivative of y with respect to x is called Leibniz's notation. This notation is useful not only because it allows us to refer to functions by using variables, but it also embodies the definition of the derivative.

So far, you only know one sure way to compute a derivative – use the definition. However, in the class activity, you began to discover some shortcuts for this process.

There are lots of shortcuts. We are going to prove one of them for you so you can see what a proof might look like. First, if we want to prove something about derivatives, the only tool we have is the definition. (*In fact, mathematical definitions are often devised to give you the power to prove things. That is why you need to understand every detail of each mathematical definition. And that is why your instructor is so picky about them.*). Now, let's use the definition of the derivative to prove the following shortcut, called the **power rule**:

Theorem: (The Power Rule). The derivative of $f(x) = x^n$ is $f'(x) = nx^{n-1}$, for any whole number n. Using Leibniz's notation: $\frac{d}{dx}(x^n) = nx^{n-1}$.

To prove this theorem, we can start by letting $f(x) = x^n$. Then by the definition of the derivative,

$$f'(x) = \lim_{\Delta x \to 0} \frac{f(x + \Delta x) - f(x)}{\Delta x}$$

$$= \lim_{\Delta x \to 0} \frac{(x + \Delta x)^n - x^n}{\Delta x}$$

What are we doing here?

$$= \lim_{\Delta x \to 0} \frac{\left(x^n + n\Delta x \, x^{(n-1)} + (coeff)\Delta x^2 \, x^{(n-2)} + \cdots + (coeff)\Delta x^{(n-1)} x + \Delta x^n\right) - x^n}{\Delta x}$$

What the heck are we doing now? Make sure this makes sense to you.

Note that in the above equation, the 'coeff' just represents *whatever* coefficient turns out to be correct. It turns out that it doesn't matter what it is, so we won't bother to compute it. Let's continue with simplifying the above expression.

$$f'(x) = \lim_{\Delta x \to 0} \frac{\Delta x \left[nx^{(n-1)} + (coeff)\Delta x x^{(n-2)} + \cdots + (coeff)\Delta x^{(n-2)} x + \Delta x^{(n-1)}\right]}{\Delta x}$$

What did we do there?

Now, since $\Delta x/\Delta x = 1$, we can write

$$f'(x) = \lim_{\Delta x \to 0} \frac{nx^{(n-1)} + (coeff)\Delta x x^{(n-2)} + \cdots + (coeff)\Delta x^{(n-2)} x + \Delta x^{(n-1)}}{1}.$$

Now we'll finally take the limit and let Δx go to zero. Note that it takes all the terms down to zero except for the first term. So

$$f'(x) = nx^{(n-1)}.$$

So we have proven the power rule. *How does this compare with the conjecture you made in the class activity?*

Now this was just the case $f(x) = x^n$. What if we had a coefficient involved like $g(x) = 4x^n$ or something? *Find the derivative function of $g(x) = ax^n$ where a is just any real number constant and n is any whole number. Use the definition and good notation, and don't skip any steps.*

One can also prove that the 'derivative of a sum' is the 'sum of the derivatives' and the 'derivative of a difference' is the 'difference of the derivatives.' This is really helpful, because it lets you find the derivative of a function like $h(x) = 3x^5 + 2x^2 - 8x$ by simply taking the derivative of each term. *What is $h'(x)$?*

Lastly, when we proved the power rule, that the derivative of $f(x) = x^n$ is $f'(x) = nx^{n-1}$, we assumed n is a whole number. But what if n is not a whole number? Let's look at one useful special case, where n = ½, which gives us the square root function.

Consider $f(x) = x^{1/2} = \sqrt{x}$. Then by the definition of the derivative,

$$f'(x) = \lim_{\Delta x \to 0} \frac{f(x + \Delta x) - f(x)}{\Delta x}$$

$$= \lim_{\Delta x \to 0} \frac{\sqrt{x + \Delta x} - \sqrt{x}}{\Delta x} \qquad \text{Why?}$$

$$= \lim_{\Delta x \to 0} \frac{\sqrt{x + \Delta x} - \sqrt{x}}{\Delta x} \cdot \frac{\sqrt{x + \Delta x} + \sqrt{x}}{\sqrt{x + \Delta x} + \sqrt{x}} \qquad \text{Why did we do this?}$$

$$= \lim_{\Delta x \to 0} \frac{x + \Delta x - \sqrt{x}\sqrt{x + \Delta x} + \sqrt{x}\sqrt{x + \Delta x} - x}{\Delta x(\sqrt{x + \Delta x} + \sqrt{x})} \qquad \text{Check the algebra.}$$

$$= \lim_{\Delta x \to 0} \frac{\Delta x}{\Delta x(\sqrt{x + \Delta x} + \sqrt{x})}$$

$$= \lim_{\Delta x \to 0} \frac{1}{\sqrt{x + \Delta x} + \sqrt{x}}$$

$$= \frac{1}{\sqrt{x} + \sqrt{x}} \qquad \text{What happened to the } \Delta x?$$

$$= \frac{1}{2\sqrt{x}}$$

So we have proven that the derivative of $f(x) = \sqrt{x}$ is $f'(x) = \frac{1}{2\sqrt{x}}$. Using Leibniz's notation, $\frac{dy}{dx}(\sqrt{x}) = \frac{1}{2\sqrt{x}}$. This will certainly come in handy later. If instead of radicals, we use exponents, we have proven that the derivative of $f(x) = x^{1/2}$ is $f'(x) = \frac{1}{2}x^{-1/2}$, which is exactly what the power rule would tell us. In fact, it can be proven that the power rule works for any real exponent n.

We know from experience that the key to being able to evaluate a limit like the one above is to somehow be able to cancel the Δx. To get there, we used a technique you may have seen in your high school algebra class. We chose to multiply by $\frac{\sqrt{x+\Delta x}+\sqrt{x}}{\sqrt{x+\Delta x}+\sqrt{x}}$, which is just equivalent to multiplying by 1. Don't worry about remembering this technique, just convince yourself that it actually works.

Homework:

A teacher is one who makes himself progressively unnecessary.
Thomas Carruthers

1) Go back and answer all the italicized questions and do the italicized suggestions in the *Read and Study* section. Pay particular attention to that proof.

2) Use what you have learned above and the shortcut from the *Class Activity* to find the derivative function for the function $g(x) = 2x^5 - 4x^3 + 3x^2 + x - 5$.

3) Practice your shortcut rules for finding the formula form of the derivative of some polynomials. Start by making up your own polynomials, then, find the derivative of each. Bring your work with you to class.

4) Prove that the power rule works for $n = {}^-1$ as well by using the definition of the derivative to find the derivative of $f(x) = x^{-1} = \frac{1}{x}$.

5) Test some examples to help you to decide whether each of the following is true or false. If it is false, give a counterexample.
   a) The derivative of a sum of two functions is always the sum of the derivatives.
   b) The derivative of a difference of two functions is always the difference of the derivatives.
   c) The derivative of 'a constant times a function' is always the constant times 'the derivative of the function.'

6) Let $f(x) = x^2(x^2 - 2) = x^4 - 2x^2$.

   a) Use the shortcut rules to find the derivative function $f'(x)$.

   b) Now find the derivative of x^2 and then the derivative $x^2 - 2$. You can use the shortcuts here too. Does the derivative of a product equal the product of the derivatives? Explain using this example. Is this consistent with what you found in the Class Activity?

7) Let's continue to look at the function $f(x) = x^2(x^2 - 2) = x^4 - 2x^2$.

   a) On your calculator plot the graphs of both $f(x)$ and $f'(x)$ in the same viewing rectangle over the interval $-2.5 \leq x \leq 1.5, -10 \leq y \leq 10$, and then draw a careful sketch of both.

   b) Over which intervals does the graph of $f(x)$ appear to be rising as you move from left to right? Over which intervals does the graph of $f'(x)$ appear to be above the x-axis? What does this tell you? Does this make sense?

   c) Over which intervals does the graph of $f(x)$ appear to be falling as you move from left to right? Over which intervals does the graph of $f'(x)$ appear to be below the x-axis? What does this tell you? Does this make sense?

   d) What are the x-coordinates of all of the high points (local maximums) and low points (local minimums) of the graph of $f(x)$? For what values of x does the graph of $f'(x)$ appear to meet the x-axis? What does this tell you? Does this make sense?

   e) What is the slope of the tangent to the graph of $f(x)$ at $x = 2$?

   f) If $f(x)$ represents the volume of water in a tank at given times x, then how fast is the volume of water changing at time $x = 2$? Is the volume increasing or decreasing?

   g) What do your answers to the above questions suggest about the relationships between a function and its derivative function. Explain why these relationships make sense.

8) The concept of increasing and decreasing functions is part of the Common Core State Standards for Grade 8. Read the introduction to section 7.2.4 and do tasks 7-68 and 7-69 in the Core Connections Course 3 (grade 8) from the College Preparatory Mathematics curriculum.

   a) Answer all the questions in these two tasks.
   b) Where in these tasks are students asked to interpret, in the sense of O'Callahan's Competencies?
   c) In what ways are these tasks related to what we asked you to do in problem 7 above?

Class Activity 18:   Baseball Acceleration

Baseball is 90 percent mental. The other half is physical.
                                                                    Yogi Berra

In the Baseball Velocity class activity, we considered the situation where a pitcher throws a ball straight up from the ground, reaching a maximum height of 144 feet after 3 seconds. The algebraic formula we used to model this situation was $s(t) = {}^-16t^2 + 96t$, where s gives the position of the ball above ground in feet and t is the time in seconds after the ball is released.

In this situation, we see that not only is the position of the ball changing over time, but the velocity of the ball is also changing over time. Let's explore how the velocity changes over time.

1) Start by filling in the values for the **position** of the ball at the times shown in the table.

Time t secs	Position $s(t)$ feet	Velocity $v(t)$	Acceleration $a(t)$
0.0			
0.5			
1.0			
1.5			
2.0			
2.5			
3.0			
3.5			
4.0			
4.5			
5.0			
5.5			
6.0			

2) Now fill in the values for the instantaneous **velocity** at each time t shown in the table. What are the units for this velocity? Write down a formula for the velocity function.

3) **Acceleration** is defined as the (instantaneous) rate of change of velocity. Fill in the values for the acceleration at each time t in the table. What are the units for this acceleration? Write down a formula for the acceleration function. Explain why this acceleration function makes sense for a baseball being thrown up into the air.

4) Make a graph showing the position, velocity and acceleration functions on the same axes.

Class Activity 18B: Toying Around Part 2

Based on your graphs of the position and velocity functions, sketch graphs of the acceleration functions for each of the three toys from Toying Around Part 1.

Then write your best answers to the following:

1. What is meant by average acceleration?

2. What is meant by instantaneous acceleration?

3. How can you tell by looking at the graph of the *velocity* function whether the toy has

    a. Positive acceleration?

    b. Negative acceleration?

    c. Zero Acceleration?

4. How can you tell by looking at the graph of the *position* function whether the toy has

    a. Positive acceleration?

    b. Negative acceleration?

    c. Zero Acceleration?

5. Describe what feature of the acceleration function graph corresponds to the toy

    a. moving at a constant rate

    b. slowing down

    c. speeding up

    d. not moving

    e. turning around to move in the opposite direction

Read and Study:

> *I'm sure the manual will indicate which lever is the velocitator and which the deceleratrix.*
>
> Mr. Burns, The Simpsons

Given a function $f(x)$, its derivative $f'(x)$ is a new function that tells you the instantaneous rate of change (or the slope, if you are thinking graphically) of the original function f at x. Since $f'(x)$ is itself a function, we can also find its derivative. This new function is the derivative of the derivative of f, which we call the **second derivative** of the original function f.

The notation for the second derivative in both function and Leibniz's notation is as follows: if $y = f(x)$, then the first derivative can be denoted $\frac{dy}{dx} = f'(x)$, and the second derivative can be denoted $\frac{d}{dx}\left(\frac{dy}{dx}\right) = f''(x)$.

A great example of a second derivative is acceleration. We have already seen that if $s(t)$ is a position function, then its derivative gives the velocity of the object, and we can write $v(t) = s'(t)$. The derivative of the velocity function is the acceleration function, which is in turn the second derivative of the position function. That is $a(t) = v'(t) = s''(t)$.

In Class Activity 7 you were asked to sketch the graph of the distance traveled by a runner who starts off slowly, but quickly reaches a high speed, slows to a walk, then sprints to the finish. Suppose you came up with the graph shown here for the runner's distance traveled.

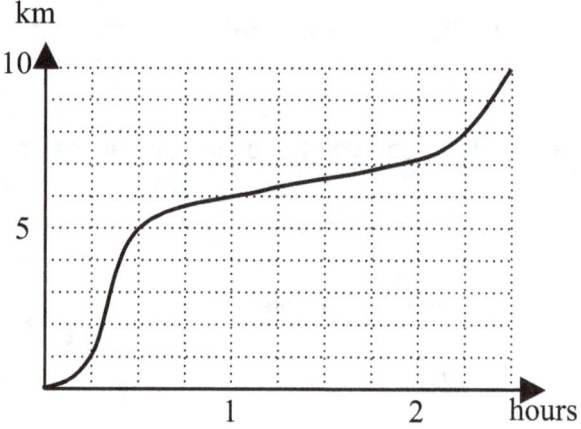

Estimate the runner's fastest velocity near the beginning of the race. Estimate at what velocity was the runner walking during the middle of the race? How fast was her sprint to the finish? Use these estimates to sketch a graph of the velocity function in the space below the position graph.

Near the beginning, as the runner is speeding up, or increasing her velocity, the acceleration is positive. When the runner's velocity is decreasing, the acceleration is negative. When the runner is traveling at a constant velocity, the acceleration is zero. *Give the approximate intervals in the graph above over which the runner's acceleration is positive, negative, and approximately zero. Use these estimates to sketch a graph of the runner's acceleration function.*

When the acceleration of an object is zero, then people usually say that the object is not accelerating, or has no acceleration. When the acceleration of an object is negative, then people often say that the object is decelerating. For example, if you are driving in your car, and you push down on the gas pedal (also called the accelerator!), your car will speed up, or accelerate. This means your car will have a positive acceleration. However, when you push on the brake, your car will slow down, or decelerate. This means your car has negative acceleration.

Note that a negative acceleration does not mean that you must have a negative velocity. For example, shown below are two graphs of the position from home for two cars on a trip. Both cars travel a total of 140 miles in 3 hours. Since both cars are always moving away from the starting point, the velocity for each car is always positive. However, in the first graph, the car has a positive acceleration and in the second graph, the car has negative acceleration. *Explain why Car A has positive acceleration and Car B has negative acceleration. What would a graph look like for a car C that also travels 140 miles away from home in three hours but with zero acceleration?*

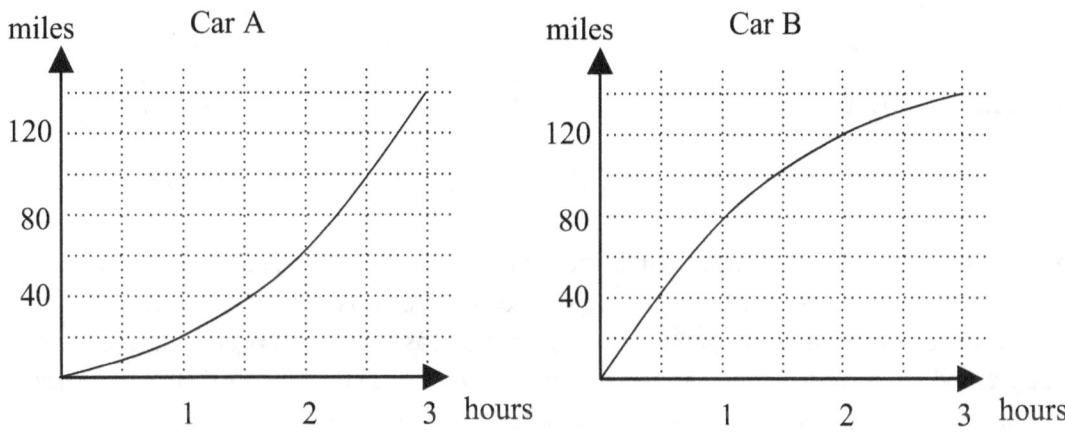

Now consider two ore cars D and E with graphs shown below. Both these cars start 140 miles away from home and travel home in three hours. *Which one has a positive acceleration? Which has a negative acceleration?*

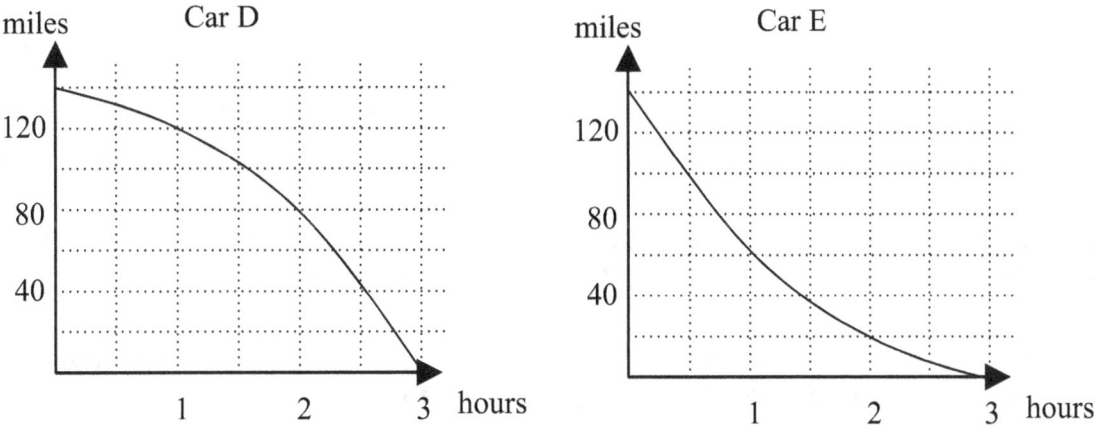

127

Look again at the four graphs above. In two of the graphs (A and E) the curve is bending up, and in two of the graphs (B and D), the curve is bending down. This property of a graph, whether it is bending up, bending down, or straight, is referred to as its **concavity.** The second derivative of a function gives information about the concavity of the graph of the function. In particular, if the second derivative is positive at a point, then the graph of the function will be bending up at that point, whereas if the second derivative is negative, the graph of the function will be bending down.

This idea of the second derivative is more common than just looking at the acceleration of objects. There are many instances when we are interested in more than just whether a quantity is increasing or decreasing. For example, a quantity that is increasing can increase a constant rate, increase at an increasing rate, or increase at a decreasing rate. Which is the case may be of extreme importance. Say a politician is talking about the national debt. Now the national debt is usually increasing. However, a politician may point out that their administration has slowed the increase of the national debt. That means that the debt is increasing at a decreasing rate. *What can you say about the derivative of the national debt with respect to time in this case? What about the second derivative?*

Connections to Teaching:

To teach is to learn twice.   Joseph Joubert

Interpreting graphs is a key component of the middle grades mathematics curriculum. As a teacher, when you are helping students to make sense of graphs, it is important that you understand the different ways a graph can increase and decrease, and use precise language to describe these behaviors and how they can be interpreted in context. *Look again at the graphs of Cars A-E. Make up 4 questions that you could ask your students about these graphs that would draw their attention to the shape of the curves and interpreting their meaning.*

Homework:

There was a footpath leading across fields to New Southgate, and I used to go there alone to watch the sunset and contemplate suicide. I did not, however, commit suicide, because I wished to know more of mathematics.
                                                                                             Bertrand Russell

1) If you haven't already done so, go back and answer all the questions posed in italics in the *Read and Study* and *Connections* sections. They are really important!

2) Estimate the acceleration at time $t = 1$ hour for cars A, B, D and E in the *Read and Study*. Don't worry about being very precise, we are just looking for ballpark figures here.

3) A poorly paced runner in a 10k race starts off slowly, but quickly reaches a high speed. She eventually tires and slows to a walk, then finally manages a sprint to the finish.
   a) Sketch a graph of the runner's total *distance* covered vs. time.
   b) Sketch a graph of the runner's *velocity* over time.
   c) Sketch a graph of the runner's *acceleration* over time.

4) Newton bakes a pie then places it on the window sill to cool. Sketch a graph of the temperature versus time. (Assume the pie was at 350 F when he took it out of the oven, and that the temperature outside is a lovely 70 F.) Sketch a graph of the derivative of the temperature of the pie with respect to time. Is the second derivative of the temperature positive, negative, or zero?

5) Explain the difference between velocity and acceleration. Give real-world examples of the following situations: (If not possible, explain why).
   a) the velocity is positive and the acceleration is negative
   b) the velocity is negative and the acceleration is positive
   c) the velocity is zero and the acceleration is non-zero
   d) the velocity is non-zero and the acceleration is zero

6) Draw a single graph of a function with the following properties:
   - $f(x)$ is always positive
   - $f'(x)$ is positive at $x = 3$ and $x = 5$
   - $f'(x)$ is negative at $x = 1$ and $x = 7$
   - $f''(x)$ is positive at $x = 1$ and $x = 3$
   - $f''(x)$ is negative at $x = 5$ and $x = 7$

7) Draw a single graph of a function with the following properties:
   - $f(x)$ is always negative
   - $f'(x)$ is positive at $x = 1$ and $x = 3$
   - $f'(x)$ is negative at $x = 5$
   - $f''(x)$ is positive at $x = 3$ and $x = 5$
   - $f''(x)$ is negative at $x = 1$

8) Read problem 8-115 in Core Connections Course 3 (Grade 8) of the College Preparatory Mathematics (CPM) curriculum.
   a) Sketch examples of the four graphs described.
   b) Describe the the first and second deriviatives the functions in each of your graphs.

9) Make a graph of the 'kid throwing the ball' function: $h(t) = -16t^2 + 32t, 0 \leq t \leq 2$. Does this graph show the path of the ball? Explain. Now make sure the velocity function v(t) = -32t + 32 makes sense with the scenario. Then use the definition of the derivative to find the acceleration function and think about that function in the context of the physical situation.

Class Activity 19: Derivatives of Sine, Cosine, and e^x

I am my own Grandpa!
Dwight B. Latham and Moe Jaffe

Never forget: a derivative function is a rate of change function; it is a *slope function*. So given the graph of a function, we challenge you to carefully sketch its derivative.

In fact, here are four functions. Working together as a group, see if you can **sketch** the graph of the derivative function in each case.

1) $f(x)$ is the function with this graph:

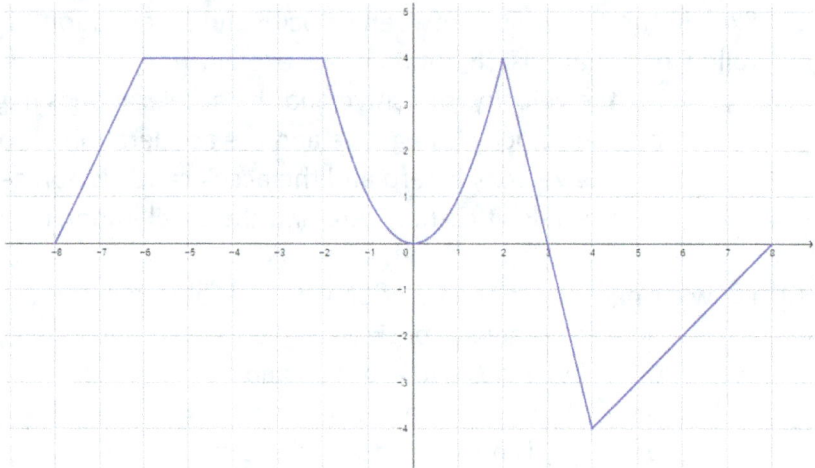

2) $k(x) = e^x$

3) $g(x) = \sin(x)$

4) $h(x) = \cos(x)$

Make a guess at an algebraic formula for each of the last three derivative functions: $g'(x), h'(x)$ and $k'(x)$.

Read and Study:

Do not worry about your difficulties in mathematics. I assure you that mine are greater.

Albert Einstein

In class we made the conjecture that the derivative of e^x is e^x by investigating the graph. Here is a proof of this conjecture using the definition of the derivative.

Theorem: If $f(x) = e^x$, then $f'(x) = e^x$. In Leibniz's notation: $\frac{d}{dx}(e^x) = e^x$.

Proof: Recall how we defined $e = \lim_{n \to \infty} \left(1 + \frac{1}{n}\right)^n$. We will re-write this formula to make it more compatible with the definition of the derivative. By making the substitution $n = \frac{1}{\Delta x}$, we get the following equivalent definition: $e = \lim_{\Delta x \to 0} (1 + \Delta x)^{\frac{1}{\Delta x}}$. Explain how we get each part of this new definition from the old one. In particular, why instead of letting $n \to \infty$ are we now letting $\Delta x \to 0$?

Now we are ready to calculate the derivative. Let $f(x) = e^x$. Then,

$f'(x) = \lim_{\Delta x \to 0} \dfrac{f(x + \Delta x) - f(x)}{\Delta x}$ (this is the definition of the derivative)

$= \lim_{\Delta x \to 0} \dfrac{e^{(x + \Delta x)} - e^x}{\Delta x}$ (substitute $f(x) = e^x$)

$= \lim_{\Delta x \to 0} \dfrac{e^x e^{\Delta x} - e^x}{\Delta x}$ Why?

$= \lim_{\Delta x \to 0} \dfrac{e^x(e^{\Delta x} - 1)}{\Delta x}$ What did we do here?

$= e^x \lim_{\Delta x \to 0} \dfrac{e^{\Delta x} - 1}{\Delta x}$ Why can we do this?

$= e^x \lim_{\Delta x \to 0} \dfrac{((1 + \Delta x)^{\frac{1}{\Delta x}})^{\Delta x} - 1}{\Delta x}$ What the heck did we do here?

$= e^x \lim_{\Delta x \to 0} \dfrac{(1 + \Delta x) - 1}{\Delta x}$ Why?

$= e^x \lim_{\Delta x \to 0} \dfrac{\Delta x}{\Delta x}$

$= e^x \lim_{\Delta x \to 0} 1$

$= e^x \cdot 1$ Why is $\lim_{\Delta x \to 0} 1 = 1$?

$= e^x$

If you can calculate a derivative, then you might be able to "un-uncalculate" a derivative, right? Like if I have a derivative function, say $g'(x) = x^2 + 3x - 7$, then *what are the possibilities for g(x)? Take a minute and figure it out.* We'll give you a hint. There's more than just one possibility: infinitely many actually. We call them all **antiderivatives** of $g'(x)$. *What would an antiderivative function tell you if the 'rate of change' function was a velocity?*

For some derivative functions, it may be a bit trickier to think up the antiderivatives. But for a few familiar functions, like $f'(x) = sin(x)$ or $h'(x) = cos(x)$, or $k'(x) = e^x$, antiderivatives are easy – now that you know some derivatives.

You discovered that the derivative of the sine function was the cosine function, so this means that an antiderivative of the cosine function is the sine function. And we note that another antiderivative of the cosine function is 'the sine function plus any constant.' *Why? What is an antiderivative of the sine function? How about e^x?*

This means that if you knew something about the rate of change of water in a tank, you could figure out water depth too – as long as you knew how deep it was to start (or at some other point in time). The depth function comes from knowing an antiderivative of the 'rate of change' of depth. The initial depth gives you the right constant. Here's an example.

The change in the water depth in a cylindrical tank is given by the function $r(t) = 3t + 1$ (ft/hour). If the tank initially contains 6 feet of water, how deep is the water after 2 hours? *Start by picturing this situation. So you see that the water will flow faster and faster into the tank?* At $t = 0$, the instantaneous flow is 1 ft/hour. At $t = 2$, the instantaneous flow is 7 ft/hour. *Estimate the water depth in the tank at the end of two hours.*

Now let's figure it out. The depth function, we'll call it $d(t)$, is the antiderivative of $r(t)$.

So $d(t) = (3/2) t^2 + t +$ some constant (we'll call it C). Check. *Do you see how we came up with it?* The thinking went something like this: Hmm. Since $r(t)$ is linear, the antiderivative is a quadratic. Now what do its coefficients have to be? A little 'guess and check' using our shortcuts and there we have it. (Helpful, huh? This is one of those times when you have to do the thinking yourself.)

Where were we? Okay, but we also know that when $t = 0$, there was 6 feet of water in the tank.

$d(0) = 6$.

$(3/2) (0)^2 + 0 + C = 6$.

So $C = 6$ and the depth function (in feet) can be given precisely as:

$d(t) = (3/2) t^2 + t + 6$.

Now all that's left to do is answer the question: how deep is the water after 2 hours?

$d(2) = (3/2)(2)^2 + 2 + 6 = 14$ feet.

Take a few minutes to study this process again and think about why it makes sense.

Homework:

In a poem the excitement has to maintain itself. I am governed by the pull of the sentence as the pull of a fabric is governed by gravity.

Marianne Moore

1) If you haven't already done so, go back and answer all the questions posed in italics in the *Read and Study* section.

2) Think about how neat it is that e^x is its own derivative. In words a high school student could understand, what does this say about the graph of the function?

3) A falling object accelerates due to gravity. The acceleration due to gravity on Earth is a constant -32 feet/sec^2. In other words, $a(t) = -32$ feet/sec^2. Note that this acceleration due to gravity is particular to Earth. On the moon, or another planet, you'd have a different constant.

   a) What is an antiderivative of $a(t)$? What does it measure?

   b) What is an antiderivative of the antiderivative of $a(t)$? What does it measure? How do we determine which antiderivative to use in each case?

   c) A ball is thrown up from the air with an initial velocity of 60 ft/sec from an initial height of 6 feet. How high will the ball go?

   d) A stomp-rocket is launched into the air from the ground and hits the ground again after 4 seconds. How high did the rocket go? What was the initial velocity of the rocket?

4) Here's another problem about tanks of water. In South Park, there are two water towers. For some reason Stan and Kyle want to drain the water out of the towers.

   a) Stan goes to the first water tower, which is full with 20,000 gallons of water. Stan flips the switch on the drain, and water pours out at a constant rate of 20 gallons per second. Find a formula for the function that gives the volume of water in the tower after *t* seconds. How long will it take for the tower to be empty?

b) Kyle goes the second water town, which is also full with 20,000 gallons of water. At this water tower, however, there is not a switch that opens the drain, but a wheel you can turn that can adjust the rate at which the water flows out. Kyle turns the wheel slowly and steadily, so that rate at which the water pours out after t seconds is given by $r(t) = 4t$ gallons per second. Find a formula for the function that gives the volume of water remaining in the tower after t seconds. How long will it take for the tower to be empty?

5) Below is a graph of a function $f(t)$ and the function $f'(t)$. Which one is which? Explain.

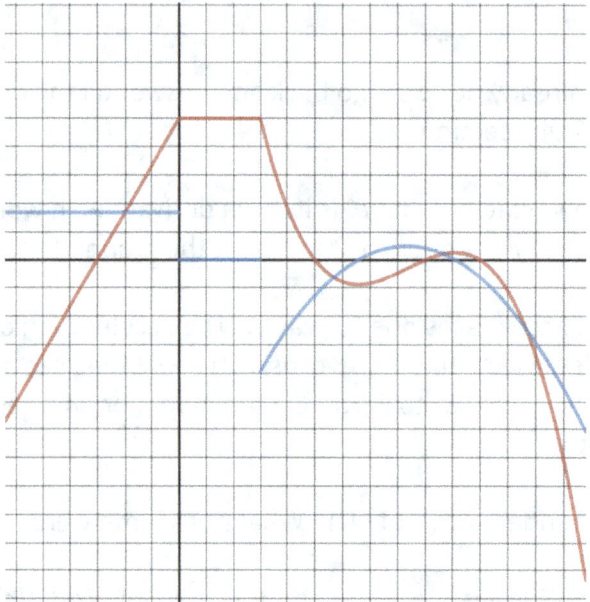

6) The rate of change of the depth of water in a tidal pool is given by the function $g(t) = 3\sin(t)$ (in feet/hour). The water is 8 feet deep at low tide. Find a formula for the depth function. In what ways does your depth function make sense, and in what ways does it fail to match what you might expect about tides?

7) Use a graph of $f(x) = \sin(2x)$ to figure out a formula for its derivative function.

Class Activity 20: Gas Mileage

Cogito Ergo Sum.

Rene Descartes

The gas mileage you get with your car depends on the speed at which you travel. Below (top) is the graph of the gas mileage as a function of speed for a test vehicle. On a recent test drive conducted by the EPA experts, data was kept on a car's speed as a function of time. It is given in the graph below (bottom).

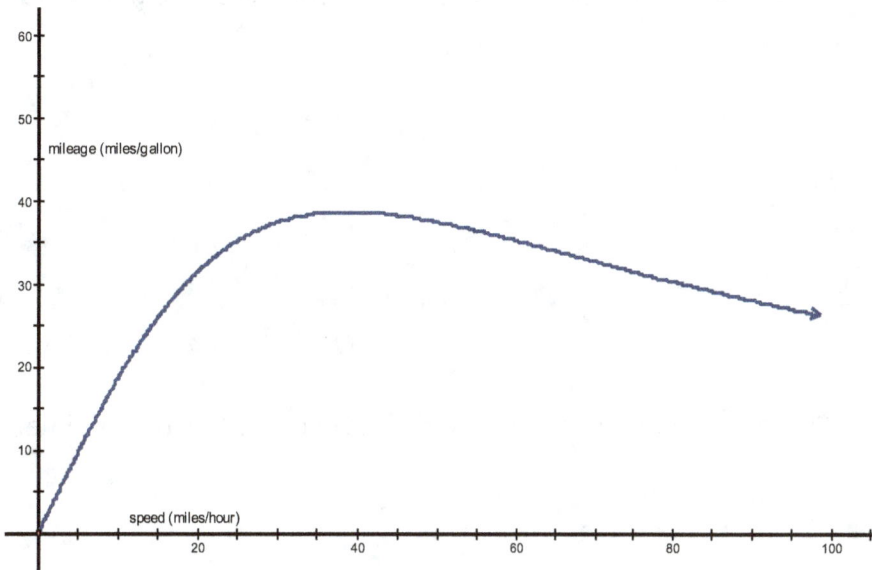

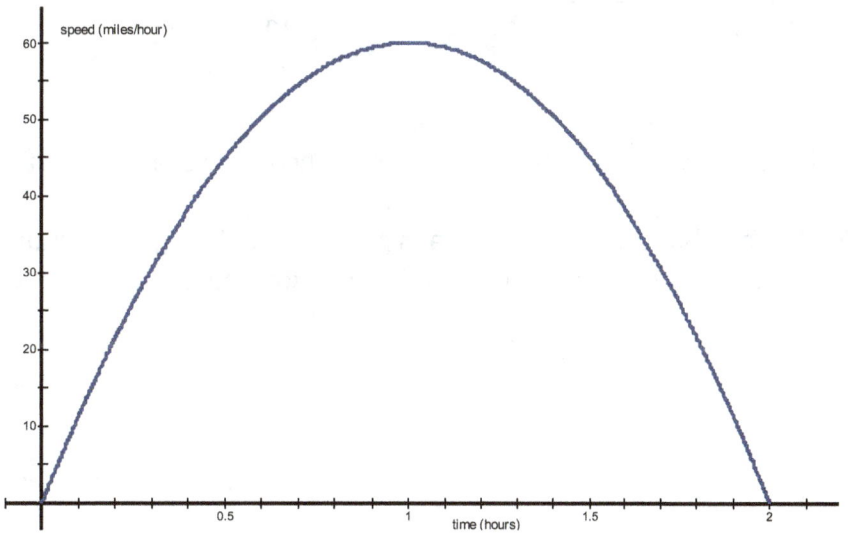

a) Does the gas mileage function seem realistic? Discuss and explain your reasoning.

b) Describe the trip the car took on the test drive.

c) Using the above graphs, sketch a careful graph of the car's gas mileage during the test drive as a function of *time*. Discuss whether the shape of this graph makes sense.

d) Find an algebraic model for the car's speed as a function of time.

e) Suppose the gas mileage function (of speed) given above is modeled by the below formula where s is in miles per hour.

$$f(s) = \frac{3000s}{1500 + s^2}$$

Use a graphing calculator to graph this function to see that it matches the graph shown.

f) Find an algebraic model for the car's gas mileage during the test drive as a function of time. Use a graphing calculator to graph this function. Does it look like your graph in part c)? Explain.

Read and Study:

The other day I got out my can-opener and was opening a can of worms when I thought: 'What am I doing?!'

Jack Handy (Saturday Night Live)

Composing two functions means putting one function inside of another. For example, if $f(x) = \sin(x)$ and $g(x) = x^3 - 7x + 3$, then $f(g(x)) = \sin(x^3 - 7x + 3))$. See how the whole function $g(x)$ is in f? Now let's think about how to build $g(f(x))$. We need to find $g(\sin(x))$. We think of g as the function that cubes *whatever you put in there*, then subtracts 7 of *whatever you put in there* and then adds 3. So $g(f(x)) = [\sin(x)]^3 - 7[\sin(x)] + 3$. You can also do things like compose a function with itself. For the above functions, $f(f(x)) = \sin(\sin(x))$. What is $g(g(x))$?

We can use function machine diagrams to help understand the composition of functions. Suppose we have a composition $y = f(g(x))$. The independent variable is x. This variable gets put into the function g, and we get some intermediate output. Let's call it u. Then this output becomes the input into the next function f which turns this into the final output y.

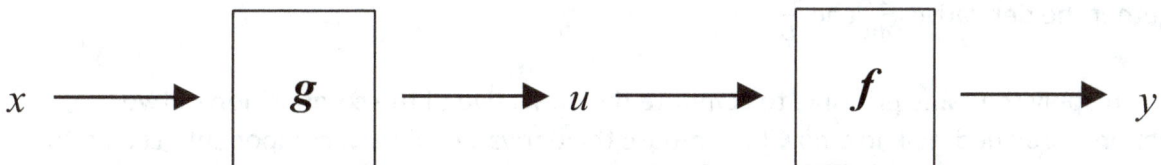

Using this intermediate variable u, we can write the two functions in this composition separately as $y = f(u)$ and $u = g(x)$. *Consider the equation $y = \sqrt{3x - 7}$. If you view this as a composition of functions, specify the two functions $f(u)$ and $g(x)$.*

Suppose we have two functions and we know how to find their derivatives. How can we find the derivative of the composition of the functions? To figure this out, we need to think about the composition of rates. Looking at the diagram, as x changes, this will result in a change in u, which in turn results in a change in y. The thus the rate of change of y with respect to x should depend on the rate of change of u with respect to x, and also on the rate of change of y with respect to u.

Consider this situation. A weather balloon is rising at a rate of 20 feet per minute. As it rises, it records the temperature, and finds that the temperature decreases at a rate of 1 degree every 100 feet. *How fast is the temperature changing with respect to time as the balloon rises? Figure this out now.*

In order to generalize this, let's analyze the situation as a composition of functions. First the independent variable is time. Given a time, we can first figure out what the height of the balloon is. Then given the height of the ballon, we can figure out the temperature. So we have two functions, a height function, and a temperature function. Using the variables in the machine diagram above, we'll let x represent the time in minutes, let u represent the height of the balloon in feet, and let. y represent temperature in degrees. Then $u = g(x)$ and $y = f(u)$.

The weather balloon rises at a rate of 20 feet per minute. This is the rate of change of height with respect to time. So $\frac{du}{dx} = 20 \frac{ft}{min}$. Since the temperature decreases 1 degree every 100 feet, we know that $\frac{dy}{du} = \frac{-1}{100} \frac{deg}{ft}$. Hopefully you were able to figure out that this means that the temperature will be decreasing at a rate of 1 degree every 5 minutes, so we have $\frac{dy}{dx} = \frac{-1}{5} \frac{deg}{min}$. Notice how we have actually just multiplied the two intermediate rates together to get the overall rate of change?

$$\frac{-1 \text{ deg}}{5 \text{ min}} = \frac{-1 \text{ deg}}{100 \text{ ft}} \cdot \frac{20 \text{ ft}}{1 \text{ min}}$$

In general, $\frac{dy}{dx} = \frac{dy}{du} \frac{du}{dx}$. This is called the **chain rule** for derivatives. It says that if you have a composition of functions $y = f(g(x))$, then its derivative $\frac{dy}{dx}$ can be found by multiplying together the derivatives $\frac{dy}{du}$ and $\frac{du}{dx}$.

We can apply the same principle to compute the derivative of the composition of two functions, provided we know how to compute the derivative of each component. Let's return to the example of the function $y = \sqrt{3x - 7}$. This function is the composition of $y = \sqrt{u}$ and $u = 3x + 7$. From previous sections, we know how to find the derivative of each of these component functions: $\frac{dy}{du} = \frac{1}{2\sqrt{u}}$ and $\frac{du}{dx} = 3$. So by using the chain rule we can compute:

$$\frac{dy}{dx} = \frac{dy}{du} \frac{du}{dx}$$

$$= \frac{1}{2\sqrt{u}} \cdot 3$$

Which we can simplify to $\frac{dy}{dx} = \frac{3}{2\sqrt{u}}$. Now looking at this answer, you might be a bit puzzled. The original function was $y = \sqrt{3x - 7}$. This is a function of the variable x. But our derivative function is a function of u. Where did this variable u come from? The fact of the matter is, we decided to introduce it in order to simplify things and to use the chain rule. The variable u represents $3x + 7$, so if we go back and remake the substitution $u = 3x - 7$ into our derivative, we will get a formula that once again is a function of x. So if $y = \sqrt{3x - 7}$, then $\frac{dy}{dx} = \frac{3}{2\sqrt{3x-7}}$.

Homework:

Self discipline is when your conscience tells you to do something and you don't talk back. — W.K. Hope

1) Suppose $f(x) = 12\sqrt{x} + 32$ is the cost in dollars of x gallons of reflective paint, and $g(x) = \frac{25}{4}x$ is the number of gallons needed to paint x miles of highway.
   a) Find a formula for $f(g(x))$. What does it represent (if anything)?
   b) Find a formula $g(f(x))$. What does it represent (if anything)?
   c) How much will it cost to paint 10 miles of highway striping?
   d) How many miles of highway can you paint for $1000?

2) Here is a picture of a function $g(x)$:
   Sketch a graph of:

   a) $g(x+3)$
   b) $g(^-x)$
   c) $g(2x)$
   d) $g(x^2)$
   e) $[g(x)]^2$
   f) $g(g(x))$

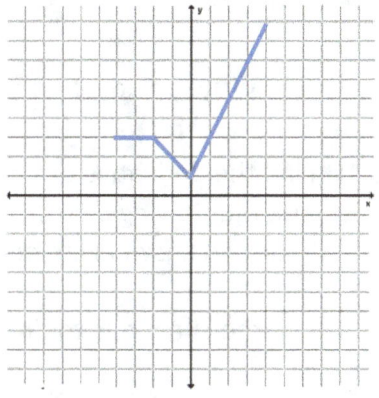

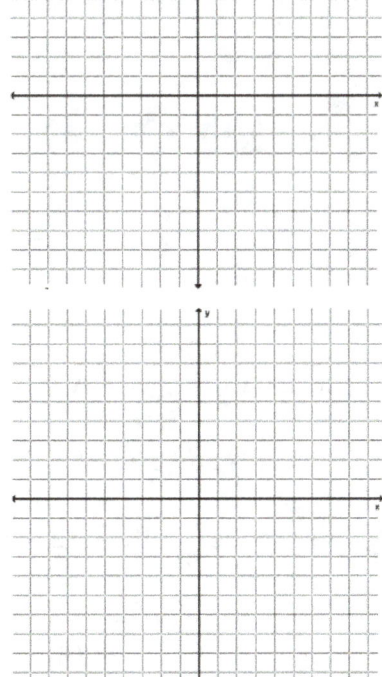

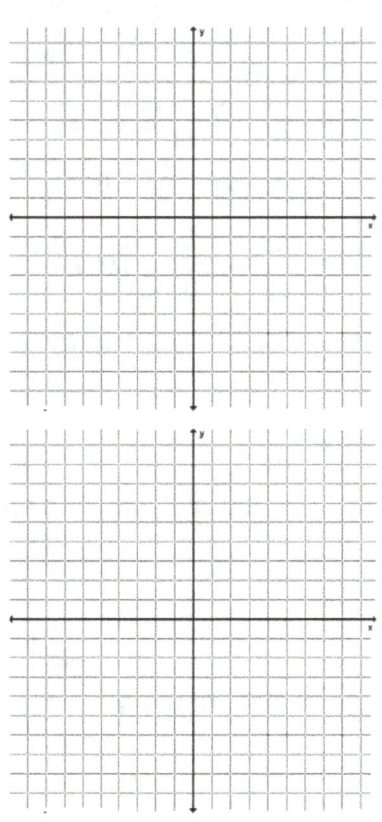

3) A new car company has developed a high-quality compact SUV that seats five people and can travel 500 miles on a single charge. The company can make these cars for $30,000, and plans on selling them for $35,000. Even at this price, demand is very high and there is a waiting list to be able to buy one of these cars. Suppose that the company can make 500 of these cars per month.

   a) At what rate is the company making money?

   b) Let y represent profit in dollars, let u represent the number of cars made (and instantly sold), and let x represent the time in months. Determine the value of the following derivatives (use proper units):

   $\dfrac{dy}{dx} = \qquad\qquad \dfrac{dy}{du} = \qquad\qquad \dfrac{du}{dx} =$

   c) Write an equation relating the three derivatives $\dfrac{dy}{dx}, \dfrac{dy}{du},$ and $\dfrac{du}{dx}$.

4) Find the derivative of $y = (2x + 5)^3$ using two methods: by using the chain rule, and by first multiplying out the product of the binomials and then taking the derivative. Show that your two answers are equivalent.

5) Use the chain rule to find the derivatives of the following functions:

   a) $y = \sqrt{8 - 3x}$

   b) $y = \dfrac{1}{x^2 - 2x + 5}$

   c) $y = e^{3x-1}$

   d) $y = e^{(x^2)}$

   e) $y = (e^x)^2$

6) The depth of water (in feet) in a tidal basin is given by the function

   $$d(t) = 4\sin(0.24\,t) + 26$$

   where t is in hours. Find an algebraic model for the rate of change of the depth of water as a function of time.

Class Activity 21: Popcorn Boxes

Extreme positions are not succeeded by moderate ones, but by contrary extreme positions.

Friedrich Nietzsche

We are going to make open-top boxes by cutting four squares off the corners of 8.5 ×11 piece of paper and folding up the sides. The plan is to find the box with the *maximum volume*. Start by making boxes where the squares you cut away have side lengths 1, 1.5, 2. 2.5 and 3 inches.

1) Make predictions. Just by looking at them, guess the order of the boxes in terms of increasing volume.

2) Now do the calculations; that is, find the volume of each of the 5 boxes. Which one has the largest volume? How good were your predictions?

3) Is there a box you haven't made yet that has the largest possible (maximum) volume? If so, find its dimensions. If not, why not?

4) Now model the volume of the box as a function of 'cut away side length' and use that model to find side length that gives the largest volume.

Read and Study:

The question is not whether we will be extremists, but what kind of extremists we will be.

Martin Luther King Jr.

This section is all about finding the biggest (maximum) or the smallest (minimum) *y*-value of a function on a given domain. These are called **extreme values**. Of course finding the extreme values of a function has practical applications. Businesses try to maximize profit and minimize loss. The Environmental Protection Agency wants to minimize CO emissions. Even nature is into extreme values. When you blow a bubble it will form into the shape that minimizes the surface area for the enclosed volume of air. Finding an extreme value for a function is called maximizing, minimizing, or in general **optimizing** a function.

How do you find the maximum or minimum value of a function? One way to estimate an extreme value is from the graph of the function. A graphing calculator can be helpful in approximating extreme values. Use a graphing calculator to graph the function $s(t) = {}^-16t^2 + 96t$. This is the function that gave the height of the ball in feet after *t* seconds in the Baseball Velocity class activity. Trace along the graph to find the maximum value for this function. *Does this give you the exact right answer? What if you zoom and trace again?* Now use the calculator's maximum finding tool (ask your instructor to show you how). *According to your calculator, what is the maximum height of the baseball? According to the calculator, at what time is this maximum height reached? Are these exact or approximate answers? Explain.*

Use a graphing calculator to graph the function $y = x^3 - x$. *Use the calculator's maximum and minimum finding tools to find the maximum and minimum values for the function $y = x^3 - x$ on the interval [-1,1]? Do you think these are exact or approximate answers?*

To calculate the exact values and locations of minima and maxima, we can use calculus. We claim that if a function is continuous on a closed interval [a, b], it must achieve both a maximum and a minimum value on that interval. *Try to draw a continuous function that doesn't.* If the function is not continuous or if the domain is an *open* interval or a half-open interval, say (a, b) or (a, b] or infinite, then all bets are off. *Try to figure out why.*

Next we claim that for a continuous function on a closed interval, the maximum and minimum values occur when the derivative is zero or undefined, or at an end point of the domain. Roughly speaking, the maximum value will occur at and endpoint or at the top of a 'hill'. Likewise, the minimum either occurs at an endpoint or at the bottom of a 'valley'. We use the terms *hill* and *valley* loosely – as metaphors rather than mathematical objects with precise definitions. The hill or valley could be rounded, in which case the derivative will be zero at the top or bottom, or it could come to a sharp point (like a V), in which case the derivative will be undefined. *Try to figure out why the derivative of a function would be undefined at a sharp point.*

To exclude functions that have sharp points like this were there derivative is undefined, we say that a function is **differentiable** if it has a derivative at every point, or in other words, if it looks like a smooth line wherever you zoom in real close.

The upshot of all this is that for functions that are continuous and differentiable, to took for minimum and maximum points, we should check to see where the derivative is zero. For example, to calculate the exact value of the maximum height of the ball for the function $s(t) = {}^-16t^2 + 96t$, we realize that at this maximum value the graph will look like the smooth top of a hill, so the derivative should be zero. This is the same as arguing that at the maximum height, the velocity of the ball should be zero. So to find this point, we can calculate $s'(t) = {}^-32t + 96$, and find where this derivative is zero. Solving $^-32t + 96 = 0$, we find that $t = 3$. So the maximum height of the ball occurs when $t = 3$ seconds. Notice that this method tells us *when* or *where* the maximum value occurs, but not the actual maximum value for the function. To find that, we can evaluate $s(3) = {}^-16(3)^2 + 96(3) = 144$ to find that the maximum height of the ball is 144 feet.

Notice how we have now come full circle with this Baseball Velocity problem. In the original class activity, our task was to come up with a function that described the height of a baseball that reached a maximum height of 144 feet after 3 seconds. From that information, we were able to find an algebraic model for the height of the baseball after *t* seconds. Now, we have just shown using calculus, that the function we came up with does indeed have a derivative of zero at $t = 3$, just as we would expect at a maximum value.

For more practice, let's look again at the function $y = x^3 - x$. We used a calculator to approximate the minimum and maximum values for this function on the interval [-1,1], but we would like to know what the exact values are. Since this is a polynomial, which we know is continuous and differentiable, we again argue that the maximum and minimum will occur when the derivative is zero. We calculate the derivative $\frac{dy}{dx} = 3x^2 - 1$, and to find were the derivative is zero, we can solve $3x^2 - 1 = 0$ to find that $x = \sqrt{\frac{1}{3}}$ or $x = {}^-\sqrt{\frac{1}{3}}$. Using what we already know about the graph of this function from our investigation earlier in this section, we know that there is a maximum value occurring at $x = {}^-\sqrt{\frac{1}{3}}$, and a minimum value occurring at $x = \sqrt{\frac{1}{3}}$. To find these maximum and minimum values, we can evaluate the function to find a maximum value of $y = \left(-\sqrt{\frac{1}{3}}\right)^3 - \left(-\sqrt{\frac{1}{3}}\right)$, and a minimum value of $y = \left(\sqrt{\frac{1}{3}}\right)^3 - \left(\sqrt{\frac{1}{3}}\right)$.

Show that the maximum value can be written as $\frac{2}{3\sqrt{3}}$ and the minimum value can be written as $\frac{-2}{3\sqrt{3}}$. Use a calculator to find a decimal approximation for these values, and then compare these values to your graphing calculator results for the minimum and maximum values from earlier in this section.

Homework:

One who makes no mistakes, never makes anything.
Anonymous

1) Do all the italicized things in the *Read and Study* section.

2) Sketch examples of the following, or explain why it can't be done:

   a) A continuous function that has no maximum on the domain (2, 5].
   b) A continuous function on the domain of all real numbers that has no minimum.
   c) A function on the domain [2, 6] that has no minimum and no maximum value.

3) The function $f(x) = -\frac{2}{3}x^3 - 2x^2 + 16x + 60$ from $x = 0$ to $x = 5$ represents the score on a test as a function of how many hours you study the night before. Determine the number of hours you should study to maximize your score on the test. Explain how you can calculate this *without using a calculator.*

4) A farmer has 32 meters of fencing and wants to fence off a rectangular pen for animals. One side of the pen will lie along a creek. Since cows don't like to swim, it is not necessary to fence along the creek. Find the dimensions that will maximize the enclosed area. Solve this problem two ways: using a graphing calculator and by using differential calculus.

5) A cylindrical aluminum pop can is to hold a volume of 355 ml. Find the dimensions of the can that will minimize the amount aluminum needed. How do your results compare with the dimensions of a standard sized pop can? If there is a significant difference, discuss why this might be the case.

6) On the same side of a river are two towns, Allentown and Bakersfield. Allentown is 10 miles from the river. Bakersfield is only 5 miles from the river. As measured along the river, the towns are 20 miles apart. The two towns have gotten together to build a pumping station by the river that will pump water through pipes to each of the two towns. Where should the pumping station be located in order to minimize the total length of the pipe needed?

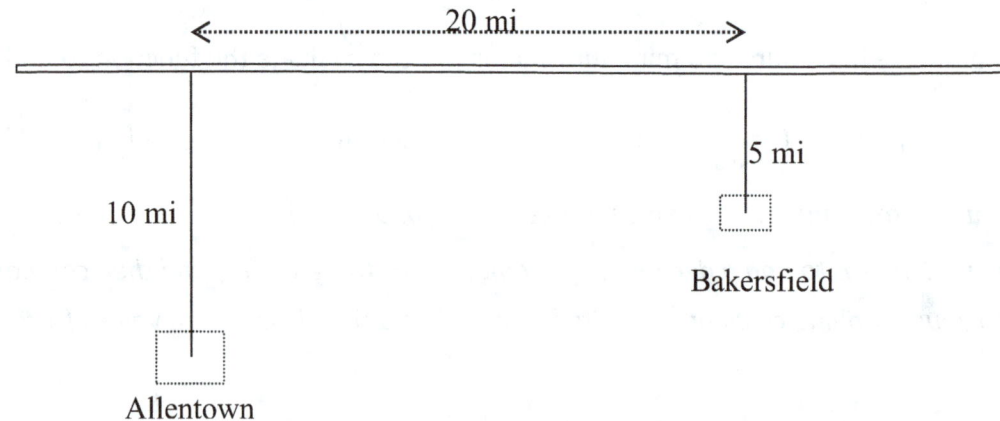

7) A town must lay a gas pipe from point A to point B. Point A is 50 feet up river from point B, and the river is about 36 feet wide. The cost for laying pipe is $7 per foot through the river and $3 per foot on dry land. What path should the town take to minimize the cost of the pipe?

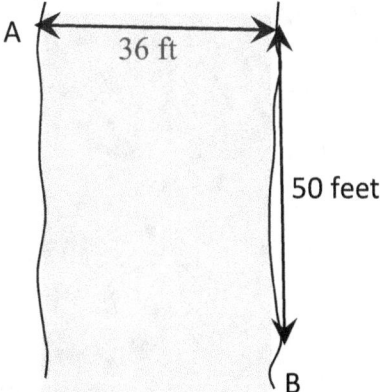

8) According to an article titled, 'Do Dogs Know Calculus?' by Timothy Pennings in (*The College Mathematics Journal*, 34(3)), a Welsh Corgi named Elvis knows how to minimize a function. When Tim (standing at A in the diagram) throws a ball into Lake Michigan at B for him to catch, Elvis (who starts next to Tim) races down the shoreline partway (say to point D) and then dives into the water, swimming for the ball. Elvis is trying to *minimize his time to the ball* – and he runs faster than he swims. And he is remarkably good at taking the theoretically correct path.

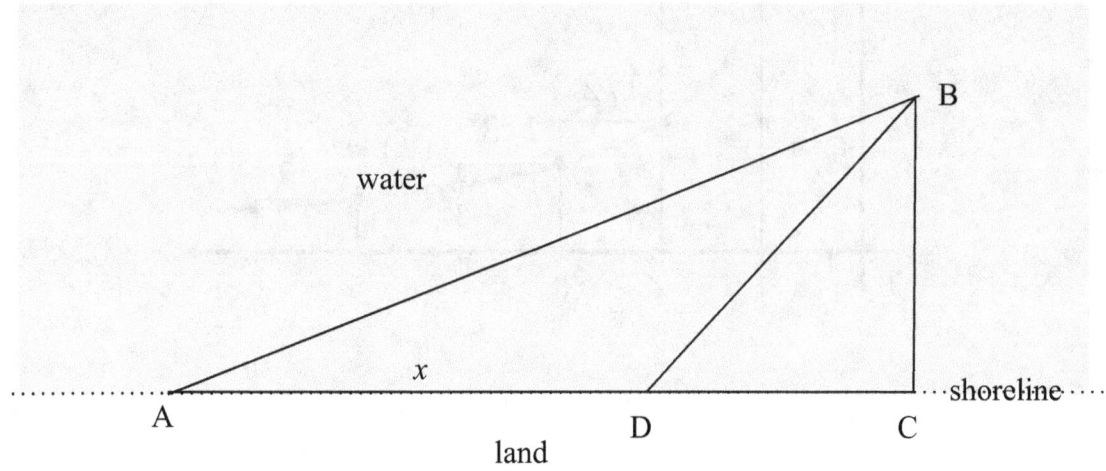

Here are the facts: Elvis runs at a rate of 6 meters/second. He swims at a rate of 1 meter/second.

If the ball is thrown to a point with coordinates (15, 9), that's 15 meters straight down the shore and 9 meters straight into the water, how far will Elvis run along the shore before diving into the water?

Chapter Four: The Definite Integral

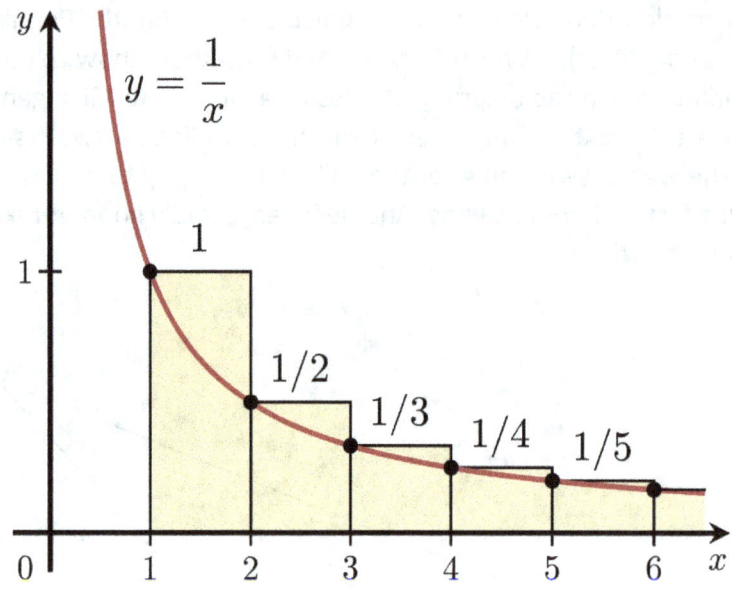

Class Activity 22: Speed Racers

It is not once nor twice but times without number that the same ideas make their appearance in the world.
　　　　　　　　　　　　　　　　　　　　　　　　　　Aristotle

1) The graphs below show the *velocities* of two cars as they race along a straight road. Car A started out with an extremely high acceleration, then eased up a little. Car B started very slowly but then and then sped up quickly. At point P, the two cars had the same velocity.

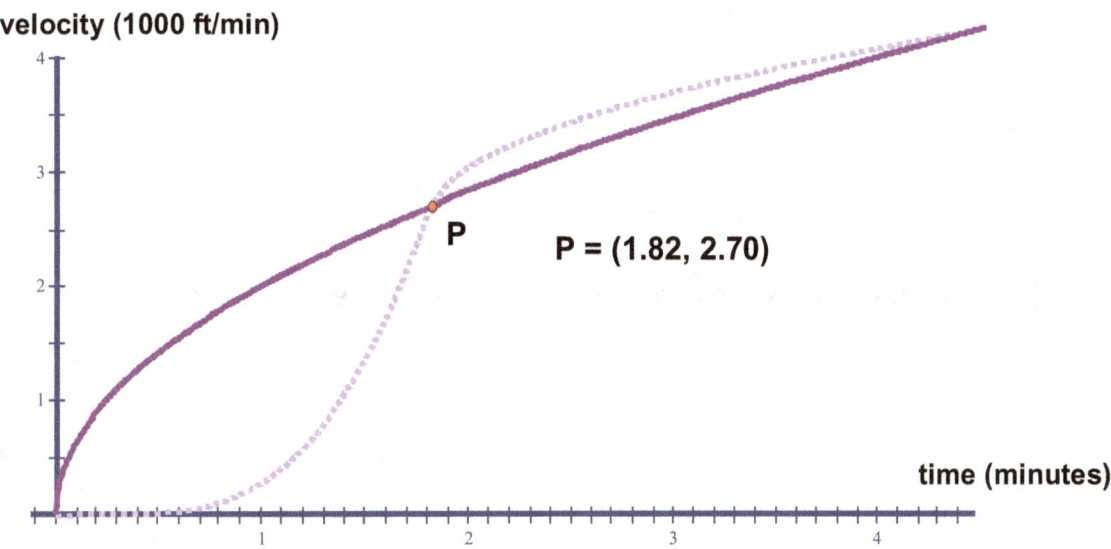

P = (1.82, 2.70)

a) Why do the graphs look curved if the road is straight?

b) Which car was ahead after 2 minutes. How do you know?

c) After four minutes, one of the cars crossed the finish line and won the race. Which car won? How do you know?

d) Estimate the length of the race. Explain how you determined this.

2) Suppose that the velocity of an accelerating car is measured as it travels down a highway, and the following table is produced:

Time (seconds)	0	2	4	6
Velocity (ft/sec)	10	24	30	34

You need to find a reasonable approximation for how far the car has traveled over the six-second interval. (No – you can't just look at the odometer.)

a) What is the greatest distance the car could have traveled? Why?

b) What is the shortest distance the car could have traveled? Why?

c) What is your best approximation for the distance the car traveled? Explain how you determined this approximation is "best."

Read and Study:

Mortals, congratulate yourselves that so great a man has lived for the honor of the human race.

Epitaph of Sir Issac Newton

We are working up to another big idea of calculus: the definite integral. We continue the development of this idea using what you learned from the Speed Racers activity.

Suppose that velocity data for the car in problem 2 is taken at one-second intervals instead of two-second intervals:

Time (seconds)	0	1	2	3	4	5	6
Velocity (ft/sec)	10	20	24	27	30	32	34

Take a minute to compute your best upper and lower bounds on the distance the car has traveled during the 6-second interval. We see that these bounds can be shown geometrically as the sum of the areas of a bunch of rectangles. In the case of an increasing function, the "right hand" rectangles (*why do we call them this?*) give the upper bound and the "left hand" rectangles give the lower bound.

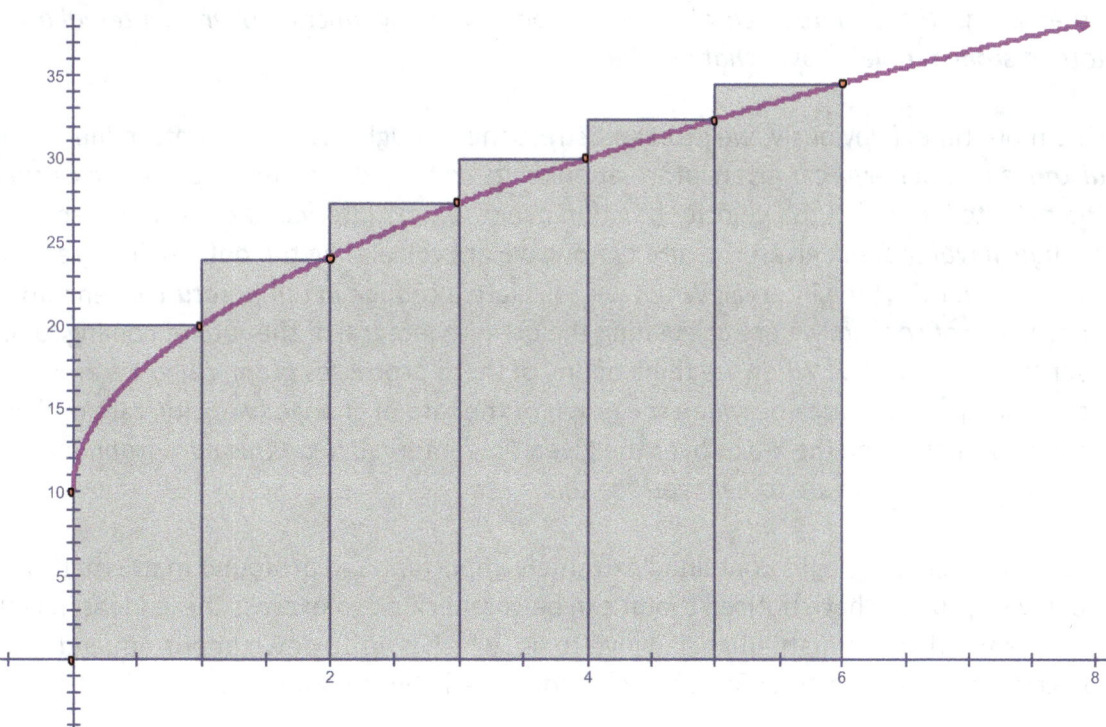

149

Suppose we tell you that a model for the car's velocity is $v(t) = 10t^{1/2} + 10$. With this new information, can you now find an even better approximation for the distance traveled? Sure. You could take data every ½-second or every ¼-second ... hmm sounds like a limit is coming.

Imagine the pictures of the rectangles for the half-second data. Sketch them in. Do you see there would be twice as many? Do you see that they would give a better approximation for the distance traveled? In general, the more rectangles, the better the approximation. *Make sure you understand this.*

The *exact* distance traveled will be the area under graph of the velocity function from *t*=0 to *t*=6. When you compute upper and lower bounds for the distance, you are simply approximating that area using a sum of rectangles. The exact area is actually the limit of the right hand (or left hand) sums as you let the width of the rectangles go to zero, or thought of another way, as you let the number of rectangles increase without bound (go to infinity). This limit is called the definite integral. The **definite integral** is defined as the limit (as the rectangle widths go to zero) of the sum of the approximating rectangle areas. If the function is nice and continuous, it doesn't matter if you think of right-hand rectangles or left-hand rectangles. The limit will turn out the same in either case.

We have just figured out something really neat and really important. When we have a function that gives the velocity of a car, to find out the total distance traveled by the car, we compute the area under the graph of the function, which is called the definite integral of the function. In other words, *the definite integral, or area under a **velocit**y function over an interval gives the total **distance** traveled over that interval.*

One more time. (Obviously, we just can't stress this enough.) When we determine *the total distance traveled* by a car given information on its *velocity* during the trip, we are calculating the definite integral of the velocity function over the time interval of the trip. When we find the *change in water depth* given the *rate of flow*, we are calculating the definite integral of the flow function over the time interval. When we calculate a *change in temperature* given information on the *rate of cooling*, we are calculating the definite integral of the rate of cooling function over the time interval. When we think of any of these processes graphically, we are determining the net area between the graph of the rate of change (velocity, rate of flow, rate of cooling) function and the *x*-axis over the given time interval. *Read this paragraph one more time and sketch a picture to help you see this.*

The preceding paragraphs contained extremely important and profound mathematical ideas. However, you may have noticed that it can be complicated to express these ideas in words. Now we will do what mathematician love to do, which is write down important and profound ideas in a compact, abstract form by using some well-chosen notation.

First we introduce some notation for the definite integral. Let's return to the case of finding the total distance traveled over the first six seconds for the car traveling with velocity $v(t) = 10t^{1/2} + 10$. We have already determined that this total distance traveled is the **area under the graph** or the **definite integral** of the function $v(t) = 10t^{1/2} + 10$ from $t = 0$ to $t = 6$. We denote this definite integral as $\int_0^6 (10t^{1/2} + 10) dt$.

We know this looks weird. We interpret this notation like this: "Sum areas of rectangles of height $10t^{1/2} + 10$ and width dt (really, super-duper small width, in other words) between $t = 0$ and $t = 6$." The integral symbol is an elongated "S" to help you remember that you are finding a *sum*. We call zero the **lower limit** of the integral and six is the **upper limit**. The function $v(t) = 10t^{1/2} + 10$ is called the **integrand** in this integral.

In general,

$$\int_a^b f(x) dx$$

denotes the **definite integral** of a function $f(x)$ from $x = a$ to $x = b$, which represents the area under the graph of $f(x)$ over the interval $[a, b]$.

The idea of the definite integral is another huge idea in mathematics, right up there with the idea of the derivative. Here are these two big ideas stated together for contrast:

1) When you start with a 'quantity' function, $f(x)$, its derivative, or 'rate of change' function $f'(x)$ shows up as the slope of the tangent line to the graph of $f(x)$. This big idea can be written down symbolically by the formula

$$f'(x) = \lim_{\Delta x \to 0} \frac{f(x + \Delta x) - f(x)}{\Delta x}.$$

2) When you start with a 'rate of change' function, $f'(x)$, the change in the 'quantity' function, $f(x)$, shows up graphically as the accumulating (signed) area between the graph of $f'(x)$ and the x-axis. This big idea can be written down symbolically by the formula

$$\int_a^b f'(x) dx = f(b) - f(a).$$

Note that when you are given a function, you will need to pay careful attention to whether it represents a 'quantity' or a 'rate of change.' *Read these again and again and think about them until you are sure you understand them. Try hard to understand the relationship between that symbolic formula and the idea written in words.*

Homework:

> *Nature laughs at the difficulties of integration.*
> *Pierre-Simon*

1) In the second speed racers activity, suppose you were given a model for the car's velocity as $v(t) = 10t^{1/2} + 10$. On your calculator plot this function on the time interval [0, 6] and then draw a careful sketch.

   a) Calculate the area between the curve and the x-axis between $t = 0$ and $t = 6$, using the average of right-hand and left-hand sums with six rectangles. (If you already did this – good for you. You are using the *Read and Study* section as you should.)

   b) You can find a close approximation for $\int_0^6 (10t^{1/2} + 10)dt$ on your graphing calculator, if you graph the integrand and use the "area" or "definite integral" command. Ask your instructor to show you how to do this.

   c) To find the *exact* value for $\int_0^6 (10t^{1/2} + 10)dt$, figure out the formula for the position of the car at time t, and use this formula to figure out the total distance traveled over the first six seconds. Compare this exact answer to your approximations for the area in parts a) and b).

2) Spend fifteen minutes practicing explaining the idea of the definite integral. Can you draw a picture to illustrate how a limit comes into play?

3) Find the distance the test vehicle was driven on the test drive in Class Activity 19.

4) Find the exact value of $\int_2^5 (12 - 2x)dx$ two different ways. (A graphing calculator can not be one of them, even though it may get lucky and give you an exact value).

5) Can the definite integral turn out to be a negative quantity? Give an example of a function on an interval that has a negative definite integral or argue why such a function cannot exist.

6) The graph below shows the velocity of a hiker in miles per hour on a long hike.

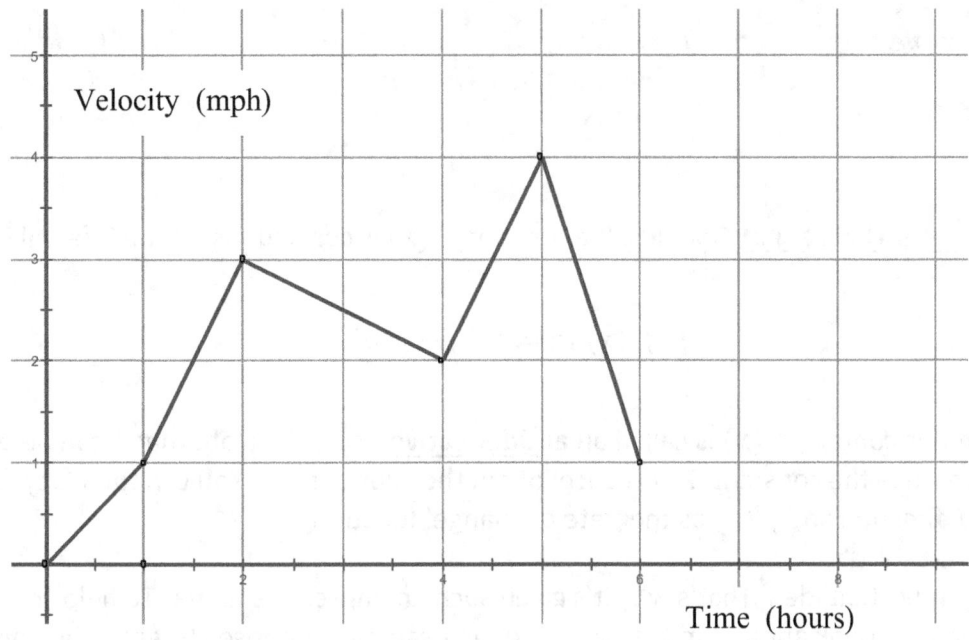

a) Compute $\int_2^4 v(t)dt$. Interpret the meaning of this value.

b) How long was the hike?

c) Was the hiker's velocity ever negative? If so, when?

d) Was the hiker's acceleration ever negative? If so, when?

e) What was the hiker's average velocity over the entire trip? How does that value relate to the graph of v(t)?

Class Activity 23: The Fundamental Theorem of Calculus

Some problems are so complex that you have to be highly intelligent and well informed just to be undecided about them.

Laurence J. Peter

The Fundamental Theorem of Calculus: If a function $f'(x)$ is continuous on the interval $[a, b]$ then

$$\int_a^b f'(x)dx = f(b) - f(a).$$

Recall that in this context, $f(x)$ is called an **antiderivative** of $f'(x)$. (Note that f can be *any* antiderivative since the constant C will cancel out in the subtraction.) You can think of $f(x)$ as the 'quantity' function and $f'(x)$ as the 'rate of change' function.

This is a very important idea. That's why it's given such an impressive name! To help you understand it, let's think about it in the context of an example. Suppose *v(t)* gives the volume of water in a tank at time *t* and suppose that *r(t)* is the rate at which water is entering the tank. (A negative rate means water is leaving the tank.) Both functions are sketched below:

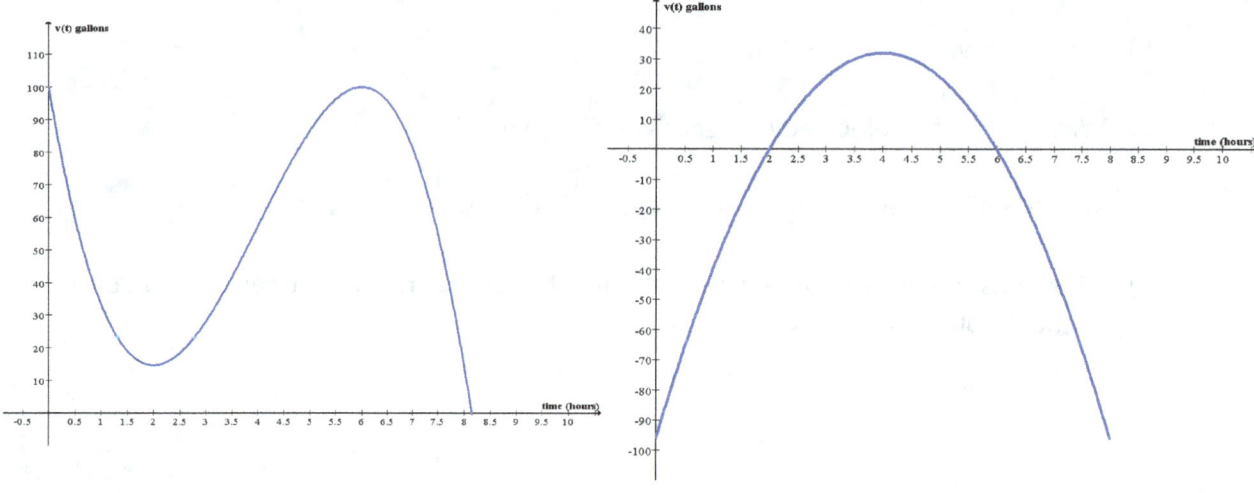

1) What is the change in volume of the tank from *t* = 2 to *t* = 4 hours? (Write the *expression* that yields the answer as well as the answer.) From *t* = 0 to *t* = 8 hours? Draw a careful picture to show how each of the above quantities can be seen on the graph of *v(t)*.

2) Now by just using the graph of $r(t)$ can you estimate the change in the volume of the tank was from $t = 2$ to $t = 4$ hours? From $t = 0$ to $t = 8$ hours? Explain how to do this, and then discuss how this relates to the fundamental theorem. Make sure everyone in your group understands this. Draw a careful picture to show how each of the above quantities can be seen on the graph of $r(t)$.

3) Now find algebraic models for these functions and use them to answer the above questions. How are these two functions related? Can you use your models to help you make sense of the Fundamental Theorem of Calculus? Fight to understand this!

Read and Study:

Sir Isaac Newton

Nature and Nature's Laws lay hid in Night.
God said, "Let Newton be" and all was light.
Alexander Pope

Leibniz hoped that mankind could rid everyday
life of its pervasive imprecision and irrationality.
William Dunham
Journey Through Genius

Gottfried Wilhelm Leibniz

(Source of photographs: http://www-history.mcs.st-and.ac.uk/history/Mathematicians/)

The Fundamental Theorem of Calculus says that if a derivative function $f'(x)$ is continuous on the interval [a, b] then $\int_a^b f'(x)dx = f(b) - f(a)$. In words, the definite integral of the derivative of a continuous function over an interval gives the total change in that function over that interval. The reason why this theorem is so profound is that it relates two fundamental ideas in calculus: the idea of the derivative and the idea of the definite integral.

While we can not offer you a rigorous proof of this theorem, we have now explored two examples in some depth that illustrate this idea: the speed racer car in the previous section and the water in the tank problem in the class activity of this section. Hopefully in these examples you were able to see why finding the area under the rate of change function gives you the total change in the original function.

We can also offer you an intuitive explanation for why this theorem makes sense. Consider the definite integral $\int_a^b f'(x)dx$. This represents the area under the graph of the function $f'(x)$ over the interval from $x = a$ to $x = b$. As we have seen, to compute this area, we can think of chopping this interval up into many infinitely small intervals of length dx, and filling up this area will infinitely many rectangles. The heights of each of these rectangles is given by $f'(x)$ and the widths given by dx, so the area of each rectangle can be represented by the product $f'(x)dx$, and we view the integral $\int_a^b f'(x)dx$ as the result of summing up the areas of all of these rectangles.

Now remember that $f'(x)$ is a derivative function. It is the derivative of the function $y = f(x)$. Using Leibniz's notation, we can write $f'(x) = \frac{dy}{dx}$ and think of the derivative as a slope $\frac{dy}{dx}$, where dy is the change in the function $y = f(x)$ over an infinitely small interval dx. Making this substitution into the integral we get $\int_a^b f'(x)dx = \int_a^b \frac{dy}{dx} dx$. Now if we pretend we can manipulate this infinitely small quantify dx as if it were finite, we could simplify the integral to $\int_a^b \frac{dy}{dx} dx = \int_a^b dy$. Now we interpret this last form to mean sum up all of the infinitely small changes in the y-value of the function as x goes from a to b. But if we *add up all of the changes in y* over the interval [*a,b*], we would just get the *total change in the function*. In other words, we should get $f(b) - f(a)$.

You now have seen all of the big ideas of calculus:

- **Limit**: A mathematically precise way to think about the idea of getting arbitrarily "close" or the idea of growing "without bound."

- **Continuity**: The idea of a graph being unbroken. This one is intuitively easy.

- **Derivative**: The derivative is the instantaneous 'rate of change' of a function. The derivative shows up as the slope of the tangent line to the curve of the 'quantity' function. A function is **differentiable** if it has a derivative at every point – or, in other words, it looks like a line wherever you zoom in real close.

- **Definite Integral**: The definite integral measures the net area between the graph of the 'rate of change' function and the *x*-axis.

- **Fundamental Theorem of Calculus**: This theorem states a relationship between the derivative and the definite integral. *Try to state this relationship in your own words.*

This collection of ideas constitutes one of the great achievements of the human race: The Calculus. Wow. What a statement. While calculus has been under construction since the time of Archimedes (287-212 B.C.), two men are credited with simultaneously creating it: Sir Issac Newton (1642-1727) and Gottfried Wilhelm Leibniz (1646-1716). Newton recognized that the idea of limit was fundamental to understanding tangents and areas, and his book *Principia Mathematica* (1687) is arguably the most important mathematical work ever written. However, we owe our modern notation and the Fundamental Theorem of Calculus primarily to Leibniz.

Think these two were friends? Nope. In fact there was quite a rift between them. See, Leibniz was the first to publish the calculus – and then the British asserted that Leibniz, who had seen some of Newton's letters, had stolen the ideas. Today, historians conclude that it is likely that both men independently arrived at the big ideas, and thus both are credited with the invention of the calculus.

These are the same ideas that students study in a traditional two-semester calculus course. The main difference between a traditional course and yours is that in a traditional calculus sequence, students learn *lots* of applications and techniques for solving specific classes of problems. We skipped all that because our purpose was not to make you engineers or other 'users of calculus.' Instead, we wanted to make you aware and appreciative of the fundamental mathematical ideas that pervade the school curriculum. When you teach your students about models, motion, and rates of change, we want you to understand where those ideas are headed.

Homework:

Whatever you do or dream you can, begin it. Boldness has genius, power and magic in it. Begin it now.

Goethe

1) Explain each of the big ideas of the calculus in your own words.

2) Harry Covaire takes a bike ride. The velocity of his bicycle in miles per hour on a straight road is given by the function *v(t)* shown in the graph below over the time interval from 0 to 3 hours. The bike was 20 miles from home at time *t* = 2.5 hours.

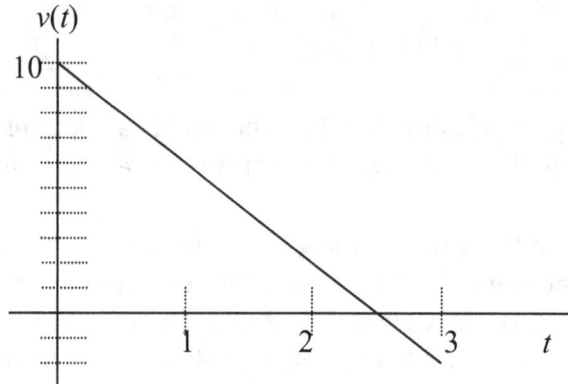

   a) Find an algebraic model *s(t)* for the position of the bike (distance from home) as a function of time.
   b) Sketch a graph of the position function *s(t)* over the time interval from *t* = 0 to *t* = 3 hours. Make an accurate scale on the *y* axis.
   c) Compute $\int_1^2 v(t)dt$. Interpret the meaning of this value in the context of the bike ride.

   Explain how you can find this value using the graph of *v(t)*, and how you can find this value using the graph of *s(t)*.

d) Compute $\int_{2}^{3} v(t)dt$. Interpret this value in the context of the bike ride.

e) Find a formula for $v'(t)$. What does $v'(t)$ represent in the context of the bike ride?

3) Use the graphs from your class activity to answer the following questions. Recall that $v(t)$ shows the volume of water in a pool as a function of time, and $r(t)$ shows the rate at which water is pumped in (and out) of the pool.

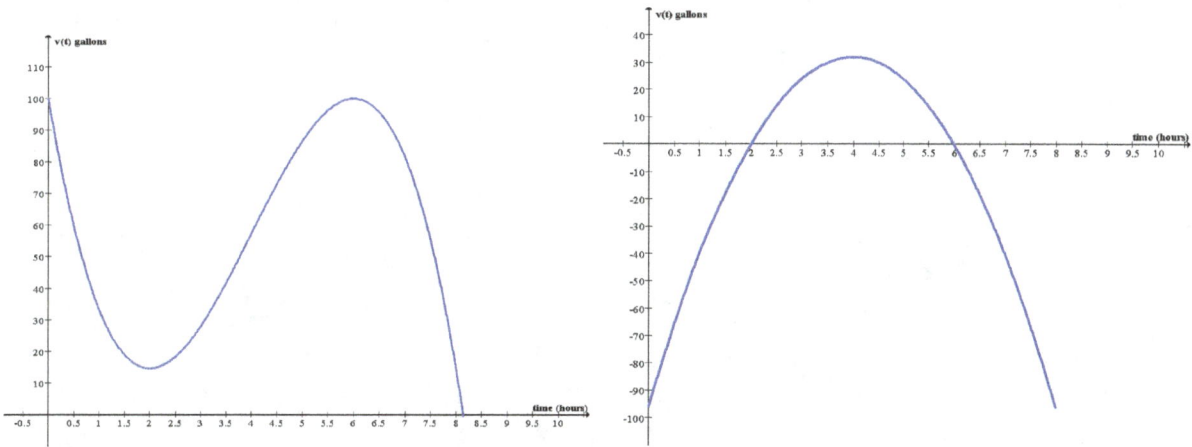

a) Write an expression for the average rate at which water enters the pool from $t = 2$ hours to $t = 7$ hours in terms of $v(t)$ and then in terms of $r(t)$. Draw a careful picture to show how each of the above quantities can be seen on the graph. Use your graphs to estimate the value of each of the expressions above. Do you get approximately the same answer both ways?

b) Write an expression for the rate at which water enters the pool at $t = 3$ hours in terms of $v(t)$ and then in terms of $r(t)$. Draw a careful picture to show how each of the above quantities can be seen on the graph. Use your graphs to estimate the value of each of the expressions above. Do you get approximately the same answer both times?

c) Write an expression for the average volume of water in the pool from $t = 2$ hours to $t = 7$ hours. Draw a picture to show how this quantity can be seen on the graph of $v(t)$ and estimate its value.

4) The velocity of a car on a straight road is given by the function $v(t) = {}^-6t + 7$. If the car was 12 miles north of home at time $t = 2$ hours, find a model $d(t)$ for the distance the car is from home as a function of time. How does the distance the car travel during the first 2 hours ($t = 0$ to $t = 2$) show up on a graph of $v(t)$? How does it show up on a graph of $d(t)$?

5) A cup of coffee at 90º C is put into a 20º C room when $t = 0$. If the coffee's temperature is changing at a rate given in 'ºC per minute' by $r(t) = -7e^{-0.1t}$, with t in minutes, estimate, to one decimal place, the coffee's temperature when $t = 10$.

7) What is the definite integral of the sine function between $x = 0$ and $x = 2\pi$? Using our new notation, this would be written like this: $\int_0^{2\pi}(\sin x)dx$. This value can be calculated in four different ways, see if you can think of all four.

Class Activity 24: A Pipeline Problem

For the things of this world cannot be made known without a knowledge of mathematics.

Roger Bacon

You now work for a petroleum company called *Oil is Us*. Your company has decided to expand its drilling in the Alaskan wilderness and you are charged with determining the most cost-effective pipeline route in connecting two wells in the area (ignoring any environmental consequences of your route). The map below shows the relevant land, with the wetland area marked. An existing oil well is located approximately at the point labeled B. The new well is to be dug at point A. You may assume both A and B are adjacent to the swamp. Your team must decide where the pipeline installation company is to lay connecting pipe. Here are the facts:

- Straight pipe must be used at a cost of $1.80/foot.
- Only one elbow joint may be used. Assume that the elbow joints may be made with any angle measure.
- Installation cost is $2/foot on dry land and $6/foot in the swamp.

Determine the pipeline route connecting the new well at A to the well at B for the minimum cost. Write a report to your boss describing your proposed pipeline route and convincing her that it is the one that incurs the least cost. Include a discussion of alternative routes you considered and why you rejected them and be certain to describe any assumptions you made in coming to your conclusions.

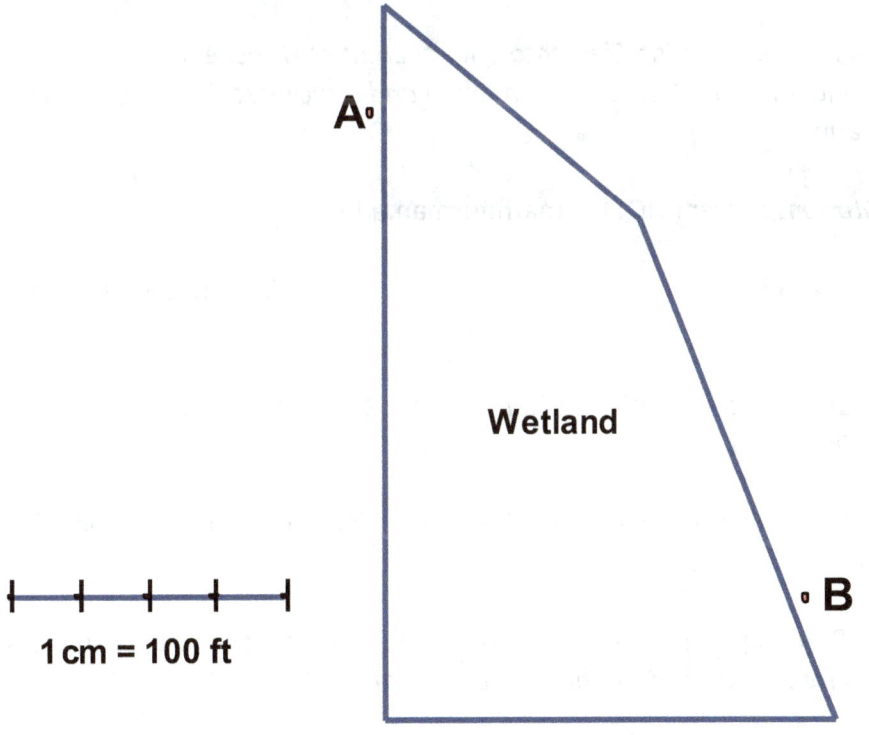

1 cm = 100 ft

References

Give credit where credit is due.
Author unknown (ironically)

- Dunham, W. (1990). *Journey Through Genius*. New York: Penguin Books.

- Halmos, P. (1985). *I Want to be a Mathematician,* Washington: MAA Spectrum, 1985.

- Hilbert, D. (2000). *The World of Mathematics*. J. Newman (Ed.)

- Gottfried Whilhem Leibniz. *New Essays Concerning Human Understanding, IV, XII*.

- Griffen, L. A., Evans, A., Timms, T., & Trowell, J. (2000). *Arkansas grade 8 benchmark exam: How do Connected Mathematics schools compare to state data?* Little Rock, AR: Arkansas State Department of Education.

- Lappan, Fey, Fitzgerald, Friel, & Phillips. (1998). *Connected Mathematics*. Dale Seymour Publications.

- Lloyd, G. & M. Wilson. (1998). Supporting innovation: the impact of a teacher's conceptions of functions on his implementation of a reform curriculum, *Journal for Research in Mathematics Education* 29(3): 248-274.

- Ma, L. (1999). *Knowing and Teaching Elementary Mathematics: Teachers' Understanding of Fundamental Mathematics in China and the United States*. Mahwah, N.J.: Lawrence Erlbaum.

- *Mathematical Quotations Server* (MQS) at math.furman.edu.

- *Mathematics in Context*. (2003). Chicago, IL: Holt, Rinehart and Winston/Encyclopedia Britannica, Inc.

- *MathScape: Seeing and Thinking Mathematically*. (1998). Columbus, Ohio: Glencoe/McGraw-Hill.

- National Council of Teachers of Mathematics (2000). *Principles and Standards for School Mathematics*. Reston, VA: Author.

- O'Callaghan, B. (1998). Computer-intensive algebra and students' conceptual knowledge of functions. *Journal for Research in Mathematical Education*, 29, 21-40.

- Pennings, T. (2003). Do dogs know calculus? *The College Mathematics Journal 34(3)*, 178-182.

- Reys, R., Reys, B., Lappen, R., Holliday, G., & Wasman, D. V. (2003). Assessing the impact of *Standards*-based middle grades mathematics curriculum materials on student achievement. *Journal for Research in Mathematics Education, 34*, 74-95.

- Riordan, J. E. & Noyce, P. E. (2001). The impact of two *Standards*-based mathematics curricula on student achievement in Massachusetts. *Journal for Research in Mathematics Education, 32*, 368-398.

- Smullyan, R. (1992). *Satan, Cantor, and Infinity and Other Mind-boggling Puzzles.* New York: Knopf.

Glossary:

The errors of definitions multiply themselves according as the reckoning proceeds; and lead men into absurdities, which at last they see but cannot avoid, without reckoning anew from the beginning.

Thomas Hobbes
in J. R. Newman (Ed.) The World of Mathematics

algebraic methods for solving equations: These are methods involving manipulating the equation to isolate the unknown using the operations of addition, subtraction, multiplication, division, and the extraction of roots.

alternating series: a sum in which every other term is negative.

antidervative: an antiderivative of a function f is any function that has f as its derivative.

average rate of change of a f (on the interval $[c, c + \Delta x]$) is $\frac{f(c+\Delta x)-f(c)}{\Delta x}$

axiom: n axiom is a mathematical statement or hypothesis that is assumed to be true without proof.

cardinality: the number of elements in a set. Two sets have the same cardinality of they can be put in a one-to-one correspondence.

coefficient: 'Coefficient' is sometimes used in place of the term *parameter*.

complement: The complement of a set A is the set of all elements that are *not* in A. (To talk about complement, you will need to know your universal set.)

complex numbers: This is the set of numbers of the form $a + bi$ where a and b are real numbers and $i = \sqrt{-1}$. This set can be modeled geometrically as the set of points in the plane and we will denote it by **C**.

composition (of functions): To compose two functions means to follow one function by another, or in an abstract sense, to put one function inside the other. The notation for f composed with g looks like this: $f \circ g = f(g(x))$

conjecture: A conjecture is a mathematical guess or hypothesis.

continuous at a point A function is continuous at $x = c$ means that $\lim_{x \to c} f(x) = f(c)$.

continuous on an interval A function is continuous on an interval if it is continuous at every point in the interval. Informally, a function is continuous on an interval if its graph has no holes, jumps, or vertical asymptotes on that interval.

converge: an infinite series is said to converge to a number N if its sequence of partial sums gets arbitrarily close to N.

converse: The statement 'If Q, then P.' is the converse of the statement 'If P, then Q.'

contrapositive: The statement 'If not Q, then not P.' is the contrapositive of the statement 'If P, then Q.'

counterexample: a counterexample is a specific example that shows a conjecture to be false.

counting numbers: this is often another name for the set of natural numbers: {1, 2, 3, 4 ...}

countably infinite: a set is countably infinite if it can be put in one-to-one correspondence with the set of natural numbers.

cubic polynomial: A cubic is a polynomial of degree three. It can be put in the form:
$$y = ax^3 + bx^2 + cx + d.$$

decimal representation (of a real number): the representation of a number as an infinite series in powers of ten.

deductive reasoning: Deductive reasoning is reasoning based on logic.

definite integral: the definite integral of a function f on an interval $[a, b]$ is defined as the limit (as the rectangle widths go to zero) of the sum of the approximating rectangle areas. If the function is continuous, it doesn't matter if you think of right-hand rectangles or left-hand rectangles. The limit will turn out the same in either case. (It is the signed area between f and the x-axis on the interval $[a, b]$.)

degree (of a polynomial): A polynomial function of *degree n* has the form:
$$y = a_n x^n + a_{n-1} x^{n-1} + \ldots + a_2 x^2 + a_1 x + a_0 \text{ with } a_n \neq 0.$$

dense: An ordered set is said to be dense if it has the property that between any two elements there is another element in that set.

dependent variable: In the context of a mathematical function, the independent variable (often denoted by x) takes on some domain of values and the dependent variable (often denoted by y) is assumed to vary with respect to changes in x.

derivative (at a point): Suppose $f(x)$ is a continuous function over an interval $a < x < b$. The derivative of $f(x)$ at the point c in that interval, which we denote $f'(c)$, is given by the following limit (if it exists):

$$\lim_{\Delta x \to 0} \frac{f(c + \Delta x) - f(c)}{\Delta x}$$

derive: In mathematics when we say we *derive* a formula, we mean that we will show, usually step-by-step, where it comes from starting from scratch.

differentiable: a function is differentiable on an interval if it has a derivative at every point on that interval. (The idea here is that if the function is nice and smooth on the interval then it is differentiable there.)

diverge: a sequence or series that does not converge is said to diverge.

domain: The domain of a function is the set of values that can be taken on by the independent variable. We think of this as the set of possible inputs for the function.

element: An element of a set is a member of that set.

empty set: The empty set is the set with no elements. (It is also called the *null set*.)

end behavior: The end behavior of a function refers to how the function behaves for very large positive or very 'large' negative values of *x*.

equation: A mathematical equation represents a balanced relationship between two expressions. One expression = another expression.

equivalent sets: Two sets A and B are equivalent if their elements can be put into one-to-one correspondence.

equal (numbers): Two numbers are equal is there is no positive distance between them on the real line.

equal sets: Two sets A and B are equal if they contain exactly the same elements. That is, every element of A is an element of B and every element of B is an element of A.

exponential change: a fixed increase in the independent variable results in a *multiplying* the dependent variable by constant factor.

exponential form: An exponential function is one having the form: $y = k b^x$ where *k* and *b* are parameters.

expression: An expression is a mathematical fragment in which no relational symbol is given. For example, '2x + 7' is an expression.

factor (of a number): A natural number *n* is a factor of a natural number *m* if there exists some natural number *k* such that *m* = *n* × *k*.

finite set: A finite set has exactly *n* elements for some whole number *n*.

function: A function is a relationship between two sets of objects so that each object in the first set is paired with exactly one object in the second set.

functional thinking: relating a value of a variable to the value of another variable

Fundamental Theorem of Algebra: Every polynomial of degree *n* has exactly *n* roots (counting multiple roots) if we allow complex numbers as roots.

The Fundamental Theorem of Calculus: If a function $f'(x)$ is continuous on the interval [*a*, *b*] then

$$\int_a^b f'(x)dx = f(b) - f(a)$$

geometric methods for solving an equation: These are methods that involve the relationships between geometric representations of the parts of the equation.

geometric series: any series of the form $a + ar + ar^2 + ar^3 + ...$ where *a* is not zero.

graph of a function: A graph of a function is the set of points (*x*,*y*) where each value of *x* in the domain is paired with its corresponding value *y* in the range.

graphic methods for solving an equation: These methods involve thinking of each side of the equation as a *function*, and finding where the graphs of those functions cross.

harmonic series: a specific series: 1 + ½ +1/3 + ¼ + ...

independent variable: In the context of a mathematical function, the independent variable (often denoted by *x*) takes on some domain of values and the dependent variable (often denoted by *y*) is assumed to vary with respect to changes in *x*.
inductive reasoning: Inductive reasoning is reasoning based on examples.

infinite discontinuity A function has an infinite discontinuity at *x* = *c* if the function grows positively or negatively without bound as *x* approaches *c*. In other words, the graph of the function has a vertical asymptote at *x* = *c*.

infinite process: a procedure that continues on forever.

infinite set: An infinite set is one that can be put in one-to-one correspondence with a proper subset of itself.

integers: The set of integers is the set: {...-3, -2, -1, 0, 1, 2, 3 ...}. It is typically denoted **Z** (for *zhalen* -- That's 'number' in German.).

interpreting: Making sense of a mathematical representation in terms of a real-life situation.

irrational numbers: This is the set of all numbers that *cannot* be written as any quotient of integers (with the denominator not equal to 0). These numbers have decimal names that neither terminate nor repeat.

jump discontinuity: A function has a jump discontinuity at $x = c$ if the function approaches one finite value from the left and a different finite value from the right as x approaches c. In other words, the graph of the function has a break in the graph at $x = c$.

limit (of a function at a point): Suppose $f(x)$ is a function defined on an interval around a point $x = c$ (but not necessarily at c itself). The *limit* of the function $f(x)$ as x approaches c is the number L if we can make $f(x)$ as close as we want to L by making x sufficiently close to c (but not equal to c). If the limit L exists, we write $\lim_{x \to c} f(x) = L$.

limit of f(x) as x approaches infinity $\lim_{x \to \infty} f(x) = L$ means that $f(x)$ is arbitrarily close to L for all x sufficiently large.

linear change: a fixed increase in the independent variable results in *adding* a constant to the dependent variable.

linear form: A line is a function of the form: $y = mx + b$ where m and b are parameters. (A linear form might also have more than one unknown like, $z = mx + ny + b$.)

logically equivalent: Two statements are logically equivalent if they are both true under the same circumstances (and both false under the same circumstances).

modeling: Modeling is finding a mathematical representation (algebraic, numeric or graphic model) which captures the essential elements of a situation.

multiple: A natural number m is a multiple of a natural number n if there exists some natural number k such that $m = n \times k$.

natural numbers: The natural numbers is the set: {1, 2, 3, 4, 5...}. It is typically denoted **N**.

null set: The null set is the set with no elements. (This set is often also called the *empty set*.)

numeric methods of solving equations: Methods that involve estimating a solution by iterative guessing and checking.

numeric table (or representation): a set of numeric data showing how specific values of an independent variable are paired with values of the dependent variable.

one-to-one correspondence: A one-to-one correspondence between two sets A and B is a pairing of the elements of the sets in which each element of A is paired with exactly one element of B and visa versa.

parameter: A parameter is a fixed constant in a mathematical expression, equation, or function.

partial sum: a sum of the first *n* (finitely many) terms of an infinite series.

polynomial: A *polynomial* is a function that can be put in the form:
$y = a_n x^n + a_{n-1} x^{n-1} + \cdots + a_2 x^2 + a_1 x + a_0$, where *n* is a whole number and $a_n \neq 0$.

power set: The power set of a set A is the set of all subsets of A. It is denoted **P(A)**.

prime number: A whole number is prime if it has exactly two distinct factors.

proof: A mathematical proof consists of a deductive argument that establishes the truth of a claim.

proof by contradiction: A method of proof by which you prove a claim by demonstrating that if the claim is not true, you get a logical contradiction.

proper subset: a proper subset of set *A* is any subset of *A* besides *A* itself.

quadratic growth. Fixed increases in the independent variable result in changes in the dependent variable that change linearly. In other words there are constant second differences in the dependent variable.

quadratic form: A quadratic is a polynomial of degree two. It has the form: $y = a(x - h)^2 + k$ or $y = ax^2 + bx + c$ (where *a, h, k* or *a, b,* and *c* are parameters).

quartic polynomial: A quartic is a polynomial of degree four.

quintic polynomial: A quintic is a polynomial of degree five.

rational functions: rational functions are quotients (ratios) of polynomials.

rational numbers: The rational numbers is the set of all numbers that can be written as a ratio of integers (where the denominator is not zero). These numbers have decimal names that either terminate or repeat. This set is typically denoted **Q** (for *quotient*).

range: The range of a function is the set of values that can be taken on by the dependent variable. We think of this as the set of outputs of the function.

real numbers: This is the set of all possible distances and their additive inverses. This set can be modeled geometrically by the set of points on a line (often called the *real line*). It is denoted **R**.

recursive thinking: relating a value of a variable to its previous value(s)

reifying: Creating a mental *object* for what was initially viewed as a *process*.

relational symbols: These symbols show a relationship between two expressions or objects. For example, <, = , and ⊇ are all relational symbols.

removable discontinuity A function has a removable discontinuity at $x = c$ if the function is undefined at $x = c$ but the limit as x approaches c exists and equals some finite number L. In this case, one could "remove" the discontinuity by defining the function to equal L at $x = c$.

root: A root of a function $y = f(x)$ is value for x that makes $y = 0$. On a graph, a root is an x-value where the function crosses the x-axis.

sequence of partial sums: every infinite series has a corresponding sequence (list) of partial sums which is found by taking the first term, sum of the first two terms, sum of the first three terms, etc...

series: an infinite number of ordered terms added together.

set: a set is a collection of definite distinguishable objects.

sine form: The sine curve is a function of the form $y = a\,sin[b(x - c)] + d$ where $a, b, c,$ and d are parameters.

slope of a line: The slope of a line is a constant value defined as the ratio of the change in y to the change in x. This value is often denoted by m in the function $y = mx + b$.

solution: A solution to an equation is any value of the unknown that makes the equation balanced. A solution to a *system of equations* is a set of values (one for each unknown) that makes all the equations balanced at the same time.

standard function forms: These are functions which have been given names and notations by the mathematical community. By changing parameters within the forms, we can use them to model problems.

subset: Set A is a subset of set B if every element of A is also in B.

tangent: (to a curve) A line that touches the curve at exactly one point and does not intersect the curve.

terms: the individual numbers (summands) in a series are called the terms.

theorem: A theorem is a mathematical statement that has been proved to be true.

translating: Making connections among mathematical representations of a problem (e.g. between a graph and a table, or an equation and a graph).

trigonometric form: A function involving sine, cosine, or tangent. For the purpose of this text our only trigonometric form will be the sine curve.

uncountably infinite: an infinite set is uncountably infinite (or uncountable) if it *cannot* be put in one-to-one correspondence with the set of natural numbers.

union: The union of two sets A and B is the set of objects that are either in A or in B or in both.

universal set: For any given problem, the universal set is the set of objects under consideration. We pretend no other objects exist.

unknown: In an equation, a value for which we want to solve is often called an unknown.

velocity: the rate of change of position. It is a signed quantity (can be positive or negative).

vertical asymptote: a function has a vertical asymptote at $x = c$ if function values approach infinity (negative or positive) as x approaches c (from either direction).

whole numbers: The whole numbers is the set: $\{0, 1, 2, 3, 4 ...\}$

y-intercept of a line: The value of the function when $x = 0$. This value is often denoted by b in the function $y = mx + b$.

www.ingramcontent.com/pod-product-compliance
Lightning Source LLC
Chambersburg PA
CBHW080543220526
45466CB00010B/3019